Somatic Cell Nuclear Transfer

ADVANCES IN EXPERIMENTAL MEDICINE AND BIOLOGY

Recent Volumes in this Series

Volume 583
TAURINE 6
Edited by Simo S. Oja and Pirjo Saransaari

Volume 584
LYMPHOCYTE SIGNAL TRANSDUCTION
Edited by Constantine Tsoukas

Volume 585
TISSUE ENGINEERING
Edited by John P. Fisher

Volume 586
CURRENT TOPICS IN COMPLEMENT
Edited by John D. Lambris

Volume 587
NEW TRENDS IN CANCER FOR THE 21ST CENTURY
Edited by Antonio Llombart-Bosch, José López-Guerrero and Vincenzo Felipo

Volume 588
HYPOXIA AND EXERCISE
Edited by Robert C. Roach, Peter D. Wagner and Peter H. Hackett

Volume 589
NEURAL CREST INDUCTION AND DIFFERENTIATION
Edited by Jean-Pierre Saint-Jeannet

Volume 590
CROSSROADS BETWEEN INNATE AND ADAPTIVE IMMUNITY
Edited by Peter D. Katsikis

Volume 591
SOMATIC CELL NUCLEAR TRANSFER
Edited by Peter Sutovsky

A Continuation Order Plan is available for this series. A continuation order will bring delivery of each new volume immediately upon publication. Volumes are billed only upon actual shipment. For further information please contact the publisher.

Somatic Cell Nuclear Transfer

Edited by

Peter Sutovsky, Ph.D.

*Department of Animal Science, University of Missouri-Columbia,
Columbia, Missouri, U.S.A.*

**Springer Science+Business Media, LLC
Landes Bioscience / Eurekah.com**

Springer Science+Business Media, LLC
Landes Bioscience / Eurekah.com

Springer Science+Business Media, LLC, 233 Spring Street, New York, New York 10013, U.S.A.
http://www.springer.com

Please address all inquiries to the Publishers:
Landes Bioscience / Eurekah.com, 1002 West Avenue, 2nd floor, Austin, Texas 78701, U.S.A.
Phone: 512/ 863 7762; FAX: 512/ 863 0081
http://www.eurekah.com
http://www.landesbioscience.com

Somatic Cell Nuclear Transfer, edited by Peter Sutovsky, Landes Bioscience / Springer Science+Business
Media, LLC dual imprint / Springer series: Advances in Experimental Medicine and Biology

ISBN: 0-387-37753-0
ISBN-13 978-1-4899-8745-7

Library of Congress Cataloging-in-Publication Data

Somatic cell nuclear transfer / edited by Peter Sutovsky.
    p. ; cm. -- (Advances in experimental medicine and biology ; v.
591)
    Includes bibliographical references and index.
    ISBN 0-387-37753-0
    1. Cloning. 2. Cell nuclei--Transplantation. 3. Somatic cells.
I. Sutovsky, Peter. II. Series.
    [DNLM: 1. Cloning, Organism--methods. 2. Cloning, Molecular
--methods. 3. Gene Transfer Techniques.   W1 AD559 v.591 2006
/ QU 450 S693 2006]
QH442.2.S66 2006
660.6'5--dc22
                                    2006023955

PREFACE

When I was first approached to edit the present volume, I was somewhat reluctant to undertake this humbling task because most of my own scientific expertise is derived from studies of mammalian fertilization rather than somatic cell nuclear transfer (SCNT). However, in the last ten years, I was very fortunate to train, collaborate, and communicate with a number of distinguished "cloners," many of whom kindly agreed to contribute to the present volume. This experience made me appreciate that the understanding of cellular and molecular events during and after SCNT is deeply rooted in our knowledge of fertilization of mammalian ovum by the speramatozoon, to which I prefer to refer to as "natural fertilization," not to say that SCNT should be regarded as "unnatural fertilization". Unique cellular events arising from SCNT have no parallel in natural fertilization and early pre-implantation development. The list includes the requirement of donor cell-nuclear remodeling, disposal of donor cell organelles, the possibility of heteroplasmy (i.e., presence of mixed mitochondrial genomes) and altered mode of oocyte activation.

Thanks to seminal contributions of my distinguished colleagues, the present volume reviews the progress of SCNT technology in major mammalian species including mouse, pig, and cattle, while at the same time, parallels can be drawn and distinctions made between the cellular and molecular basis of nuclear transfer and fertilization.

Randall Prather will review recent progress in pig SCNT and production of transgenic and knock out pigs for biomedical research and xenotransplantation, and Björn Oback with David N. Wells will summarize the advances achieved in cattle cloning. Studies of cloned mice will be revisited from a new, behavioral point of view, by a team led by Kellie L.K. Tamashiro.

To examine the cellular and molecular mechanism of SCNT, Christopher Malcuit and Rafael A. Fissore will discuss oocyte activation; Heide Schatten with Qing-Yuan Sun, and Stefan Hiendleder will review centrosomal and mitochondrial inheritance, respectively; Keith E. Latham, Shaorong Gao and Zhiming Han will address the issues of donor cell-nuclear reprogramming; Josef Fulka Jr. and Helena Fulka will assess the importance of donor cell cytoplast; and Jozef Laurincik will team up with Poul Maddox-Hyttel to address nucleologenesis and activation of ribosomal genes in cloned embryos.

Altogether, the present tome will offer readers a unique developmental, molecular, and cell-biological view of the exciting SCNT technology, while also discussing its pitfalls, detours, and possible future directions. We hope that it will help both students and experienced researchers to further develop their knowledge and understanding of mammalian embryonic development.

I would like to thank all contributors for outstanding chapters, Dr. Ron Landes and the staff of Landes Bioscience for making this unique volume possible, and Ms. Kathy Craighead for assistance with initial editing of all manuscripts.

Peter Sutovsky, Ph.D.

ABOUT THE EDITOR...

   PETER SUTOVSKY, Ph.D., is an Assistant Professor at the University of Missouri-Columbia with joint appointments in the Departments of Animal Science and Obstetrics and Gynecology. For more than a decade he has studied mammalian spermatogenesis, fertilization and pre-implantation embryonic development, with special interest in the ubiquitin system. Dr. Sutovsky's recent work described the role of proteasomal degradation of ubiquitinated proteins in epididymal sperm quality control, fertilization, pronuclear development and mitochondrial inheritance. He has published more than 75 research articles, reviews and book chapters and was recently named the inaugural recipient of the USDA-NRI Discovery Award. In high demand as a presenter, Dr. Sutovsky has been the invited plenary speaker at numerous international symposiums in Europe, South America, Asia, and the U.K. He earned his Ph.D. in Physiology from the Czech Academy of Sciences, Prague, Czech Republic in 1994 before postdoctoral appointments at the University of Wisconsin, Madison, and Oregon Health & Science University, Portland.

PARTICIPANTS

Rafael A. Fissore
Department of Veterinary
  and Animal Sciences
Paige Laboratory
University of Massachusetts
Amherst, Massachusetts
U.S.A.

Helena Fulka
Institute of Animal Production
Prague
Czech Republic
and
Czech Academy of Sciences
Institute of Experimental Medicine
Prague
Czech Republic

Josef Fulka, Jr.
Institute of Animal Production
Prague
Czech Republic

Shaorong Gao
The Fels Institute for Cancer Research
  and Molecular Biology
Temple University Medical School
Philadelphia, Pennsylvania
U.S.A.
and
National Institute of Biological
  Sciences
Zhongguancun Life Science Park
Beijing
China

Zhiming Han
The Fels Institute for Cancer Research
  and Molecular Biology
Temple University Medical School
Philadelphia, Pennsylvania
U.S.A.

Stefan Hiendleder
Department of Animal Science
The University of Adelaide,
  Roseworthy Campus
Roseworthy, South Australia
Australia

Keith E. Latham
The Fels Institute for Cancer Research
  and Molecular Biology
Department of Biochemistry
Temple University Medical School
Philadelphia, Pennsylvania
U.S.A.

Jozef Laurincik
Constantine the Philosopher
  University
Nitra
Slovak Republic

Poul Maddox-Hyttel
Department of Animal and Veterinary
  Basic Sciences
Royal Veterinary and Agricultural
  University
Frederiksberg C
Denmark

Christopher Malcuit
Department of Veterinary
  and Animal Sciences
University of Massachusetts
Amherst, Massachusetts
U.S.A.

Björn Oback
Reproductive Technologies
AgResearch Ltd.
Ruakura Research Centre
Hamilton
New Zealand

Randall S. Prather
Division of Animal Science
Food for the 21st Century
College of Food Agriculture
  and Natural Resources
University of Missouri-Columbia
Columbia, Missouri
U.S.A.

Randall R. Sakai
Neuroscience Program
Department of Psychiatry
University of Cincinnati
  College of Medicine
Cincinnati, Ohio
U.S.A.

Heide Schatten
Department of Veterinary
  Pathobiology
University of Missouri-Columbia
Columbia, Missouri
U.S.A.

Peter Sutovsky
Department of Animal Science
University of Missouri-Columbia
Columbia, Missouri
U.S.A.

Qing-Yuan Sun
Department of Veterinary
  Pathobiology
University of Missouri-Columbia
Columbia, Missouri
U.S.A.
and
State Key Laboratory of Reproductive
  Biology
Institute of Zoology
Chinese Academy of Sciences
Beijing
China

Kellie L.K. Tamashiro
Neuroscience Program
Department of Psychiatry
University of Cincinnati
  College of Medicine
Cincinnati, Ohio
U.S.A.
and
Department of Psychiatry
  and Behavioral Sciences
Johns Hopkins University
  School of Medicine
Baltimore, Maryland
U.S.A.

Teruhiko Wakayama
Institute for Biogenesis Research
University of Hawaii School
  of Medicine
Honolulu, Hawaii
U.S.A.
and
Laboratory for Genomic
  Reprogramming
Center for Developmental Biology
RIKEN Kobe
Chuo-ku, Kobe
Japan

David N. Wells
Reproductive Technologies
AgResearch Ltd.
Ruakura Research Centre
Hamilton
New Zealand

Yukiko Yamazaki
Institute for Biogenesis Research
University of Hawaii School
  of Medicine
Honolulu, Hawaii
U.S.A.

Ryuzo Yanagimachi
Institute for Biogenesis Research
University of Hawaii School
  of Medicine
Honolulu, Hawaii
U.S.A.

CONTENTS

6. NUCLEOLAR REMODELING
IN NUCLEAR TRANSFER EMBRYOS 84

Jozef Laurincik and Poul Maddox-Hyttel

7. SOMATIC CELL NUCLEAR TRANSFER (SCNT) IN MAMMALS:
THE CYTOPLAST AND ITS REPROGRAMMING
ACTIVITIES 93

Josef Fulka, Jr. and Helena Fulka

8. MITOCHONDRIAL DNA INHERITANCE AFTER SCNT 103

Stefan Hiendleder

CHAPTER 1

Nuclear Remodeling and Nuclear Reprogramming for Making Transgenic Pigs by Nuclear Transfer

Randall S. Prather*

Abstract

A better understanding of the cellular and molecular events that occur when a nucleus is transferred to the cytoplasm of an oocyte will permit the development of improved procedures for performing nuclear transfer and cloning. In some cases it appears that the gene(s) are reprogrammed, while in other cases there appears to be little effect on gene expression. Not only does the pattern of gene expression need to be reprogrammed, but other structures within the nucleus also need to be remodeled. While nuclear transfer works and transgenic and knockout animals can be created, it still is an inefficient process. However, even with the current low efficiencies this technique has proved very valuable for the production of animals that might be useful for tissue or organ transplantation to humans.

Introduction

The concept of cloning by nuclear transfer was first described by Hans Spemann.[1] Spemann was interested in the concept of nuclear equivalence. He wanted to test the theory that cells become irreversibly differentiated because they inherit an unequal amount of "nucleoplasm" and thus are not totipotent. He described a "fantastical experiment" whereby nuclei from progressively differentiating stages of embryos are transferred to the cytoplasm of enucleated oocytes. At the early cleavage stages, where the nuclei are still totipotent, it was expected that development would be recapitulated. Once the nuclei became irreversibly differentiated, i.e., lost important components of the "nucleoplasm", then development would arrest at an early stage of development. Sadly, these experiments were beyond the technical ability of the researchers of the time, and the experiments were not completed until after the death of Spemann.[2] Later Jon Gurdon and colleagues reported a number of foundational studies that evaluated how nuclei could be remodeled and reprogrammed after transfer to an oocyte.[3-5] In 1981, it was reported that similar nuclear transfer procedures could be applied to mice and result in cloned offspring from nuclei of inner cell mass cells of the blastocyst stage embryo, but not from the trophectodermal nuclei.[6] Unfortunately, these experiments were not reproducible and even led some authors to the conclusion that "nuclear reprogramming in mammals is biologically impossible".[7] This unfortunate conclusion was followed by reports, that used a different technology, of cloned sheep,[8] cattle,[9] rabbits,[10] and pigs[11] from early cleavage stage

*Randall S. Prather—Division of Animal Science, Food for the 21st Century, College of Food, Agriculture & Natural Resources, University of Missouri-Columbia, 920 East Campus Drive, E125 ASRC, Columbia, Missouri 65211-5300, U.S.A. Email: PratherR@missouri.edu

Somatic Cell Nuclear Transfer, edited by Peter Sutovsky. ©2007 Landes Bioscience and Springer Science+Business Media.

embryos. The first clones from cells cultured from embryos was in cattle,[12] and then from adult cells in sheep.[13]

These reports set the stage for the cloning of pigs from cultured and adult cells.[14] The main advantage to using cultured cells is that genetic modification can be performed on the cultured cells prior to making the animal.[15] Thus it is now possible to make very precise genetic modifications to cells in culture. Those cells with the genetic modification could then be used as donors in the cloning procedures to produce animals with that specific genetic modification: be it a gene addition or a gene knockout. The remainder of this manuscript will focus on nuclear remodeling, nuclear reprogramming, and the specific application of the cloning technology to creating transgenic pigs.

Nuclear Remodeling

When nuclei are transferred into the cytoplasm of oocytes, the degree of remodeling and reprogramming that occurs is thought to be dependent upon the state of the oocyte. If the oocyte is arrested in metaphase II of meiosis (regardless if the metaphase II chromosomes are present or not), then the transferred nucleus will undergo nuclear envelope breakdown and chromosome condensation. Such a process will result in the release of many of the proteins that are involved in regulating the 3-dimensional structure of the nucleus; e.g., nuclear lamins, small nuclear ribonucleoproteins (snRNPs), nucleolar proteins, histones, etc. If the oocyte is activated, then a pseudo-pronucleus is assembled by using proteins, RNAs, etc. from the cytoplasm of the oocyte. The reassembled nucleus is generally larger in size[16] and has a structure more similar to a zygotic pronucleus, i.e., there is a change in the types or distribution of nuclear lamins, snRNPs, nuclear matrix mitotic apparatus, nucleoli, etc.[17-21] Since structure confers function, it is thought that this exchange of proteins between the donor cell nucleus and the oocyte cytoplasm remodels the chromatin such that the nucleus is reprogrammed to behave as though it were a pronucleus.

Other structural components that regulate transcription are the histones. In mice, somatic cell type histone H1 in the donor cell nucleus is rapidly replaced by oocyte-derived H1 within 60 minutes after SCNT. The complement of nuclear histone H1 is then replaced at the two- to four-cell stage by embryo-derived H1.[22-24] This is likely a consequence of embryonic H1 gene transcription and translation, as well as of the limited half-life of oocyte-derived H1. Thus mammalian ooplasm is programmed for remodeling of the donor cell nucleus during SCNT similar to the remodeling of the sperm nucleus after natural fertilization. Rapid histone replacement after SCNT requires a highly efficient, substrate-specific proteolytic mechanism. The ubiquitin-proteasome pathway has been shown to control the turnover of fibrillarin and other major nucleolar ribonucleoproteins directly in the cell nucleus. Furthermore, both core and linker histones are common ubiquitin substrates in both somatic (reviewed in ref. 25) and germ cell[26,27] nuclei. Sperm heads also contain a certain proportion (up to 20%) of somatic cell type histones[28] that may need to be removed/replaced during natural fertilization. Thus the ubiquitin-proteasome pathway may be intimately involved with nuclear remodeling.[29]

While some components of the nucleus are remodeled, there are other components that may not be remodeled or where remodeling is not complete. The length of the telomeres provides an example of where there are conflicting results in the literature. In the sheep telomere length of the clones appears to be shorter than normal,[30] while in cattle they may be normal[31,32] or longer than normal,[33,34] and in pigs the telomere length appears to be within the normal range.[35] Clearly, more work needs to be done to better understand the regulation and complete function of telomeres.

The degree of DNA methylation is another area where conflicting results appear to be the norm. One group reported normal DNA methylation patterns in pigs,[36] but normal[37] as well as abnormal[38] patterns in cattle. They also suggest that factors associated with the oocyte chromatin can influence the methylation pattern in cloned embryos.[39]

The degree of remodeling of the proteins that associate with and alter the 3-dimensional architecture of the DNA will affect which genes are or are not expressed, as well as the epigenetic modifications that occur to the DNA. Thus the events that occur during the first cell cycle will have a major impact on reprogramming of gene expression and subsequent development.

Nuclear Reprogramming

The result of incomplete reprogramming is aberrant gene expression. A number of studies have evaluated gene expression in NT-derived embryos and have either observed correct reprogramming,[40] or incorrect reprogramming.[41,42] In studies that attempted to evaluate gene expression on a more global scale, it was found that about 95-96% of the genes were correctly expressed. Some of this 5% of the coding genes that have aberrant function have been identified.[43-45] Similarly, NT-derived mouse embryos exhibited aberrant expression in about 4% of the transcriptome as determined via microarray analysis.[46] However, it should be remembered that these measurements were taken in only the embryos that developed to the blastocyst stage. Thus, although much remodeling occurs normally after nuclear transfer, clearly in some cases it is not complete.[29]

Lee et al[47] evaluated gene expression in SCNT pig embryos and found that they had over expression of platelet-activating factor receptor. They then added platelet activating factor to the culture medium and improved development. A recent study[48] used porcine skin-derived stem cells as donors for nuclear transfer and evaluated embryos at the 8-cell stage of development by measuring Oct4, STAT3, FGFR2, Rad50, Tsg101, Bub3, Dnmt1 and histone H2a. They found that STAT3, Dnmt1 and Tsg101 message levels deviated most significantly from in vivo controls.

SCNT embryos tend to develop at lower rates than controls. Obviously part of the cause is inappropriate gene expression. We have also evaluated apoptosis in the SCNT embryos and found it to be higher than in vitro produced controls.[49] In addition, fragmentation rates were higher in the SCNT embryos compared to controls on days 4, 5 and 6.

Even though nuclear remodeling may not completely reprogram nuclei, sufficient remodeling/reprogramming does occur for supposed terminally differentiated cells such as olfactory sensory neurons to support development to term.[50,51] While these differentiated cells can be reprogrammed sufficiently to result in term development, those cells that have undergone a DNA rearrangement, such as B cells, cannot return to their original DNA arrangements even though they too can be reprogrammed sufficiently to support development.[52] Even nuclei from cells that would be considered to be cancerous can be reprogrammed to support development.[53,54]

Much of this work has been conducted in the mouse because the tools are already available. We have now developed a tool to begin to understand gene expression on a global scale in early pig embryos[55,56] and are in the process of applying that tool to nuclear transfer derived embryos that were activated by using different protocols. We expect to identify genes that are aberrantly expressed and then to be able to correlate the aberrant expression with changes in DNA methylation as described above. By applying these two tools (differential methylation hybridization and transcriptional profiling) to the same set of embryos, we hope to provide a comprehensive understanding of gene regulation in the nuclear transfer process in pigs.

Nonnuclear Remodeling

While much focus has been on nuclear remodeling there must also be changes in cytoplasmic structures so that the nuclear remodeling/reprogramming is compatible with the cytoplasm. The two structures that will be discussed here include the centrosome and the mitochondria. These two structures undergo unique morphological changes during embryogenesis.

The mitochondria have a unique embryo-type morphology and location. They assume a cortical location early and become dispersed within the blastomeres as they develop to the blastocyst stage. From the oocyte to 8-cell stage the mitochondria occur in small aggregates, and although they appear to be composed of many mitochondria, they are usually composed of only a few mitochondria with a complex structure.[57] Most interestingly they do not have cristae. It is not until the blastocyst stage that the mitochondria transform into somatic type mitochondria with numerous cristae in normal[58] and parthenogenetic[59] pig embryos. Since the mitochondrial genome encodes many proteins that must interact with proteins encoded by the nucleus, it is imperative that the two genomes be synchronized and compatible in nuclear transfer embryos. For a more in-depth discussion of nuclear mitochondria interactions in early embryos see these papers by Smith and colleagues.[60-64]

Another structure that is unique in early embryos is the centrosome. Generally the centrosome is composed of two rod shaped barrels (centrioles), each containing nine microtubule triplets in a pinwheel-like arrangement on a cross-section. The centrosome organizes the cytoplasmic microtubule cytoskeleton that facilitates chromosome migration during meiosis and mitosis and organelle positioning during interphase. The mitotic/meiotic microtubules are generally organized into a tightly focused spindle pole. However, in early embryos there are no centrioles and the spindle pole is barrel shaped, similar to plant cells where there are also no centrioles. Again it is not until the blastocyst stage that centrioles appear and the spindle becomes focused to a narrow pole.[65] So the question that arises is what happens to the centrioles from the donor cell after transfer to the oocyte? While this question has not been addressed directly, one of the molecules that associates with the centrioles is centrin. Centrin is present in germ cells, but disappears during meiosis in oocytes. Centrin is present in spermatozoa, but appears to be degraded at fertilization. Centrin reappears at the blastocyst stage. So, again we encounter the question of localization of donor cell centrin after nuclear transfer. Donor cell centrin appears to be degraded and centrin cannot be detected again until the nuclear transfer embryo reaches the blastocyst stage.[66] Thus, centrin appears to be fully reprogrammed.

The major vault protein (MVP) is a ribonucleoprotein that comprises the vault particle and is involved in multidrug resistance. During porcine embryogenesis, the MVP shows mostly cytoplasmic staining and is occasionally associated with the nuclear envelope and nucleolus precursor bodies. In nuclear transfer-derived embryos there was a greater tendency for abnormal accumulation of the MVP into patches as compared to in vivo controls, but not unlike in vitro produced controls.[67] This accumulation in the SCNT-derived embryos appeared to be dependent upon a functional proteasomal system.

Epigenetics and Large Offspring Syndrome

The abnormal phenotypes that are observed in NT-derived animals are likely traced back to errors in nuclear remodeling and nuclear reprogramming. Large offspring syndrome (LOS) was first described in cattle derived from in vitro oocyte maturation, fertilization, and culture prior to embryo transfer. Since the most common phenotype was large birth weight,[68,69] LOS has been used to describe all the abnormalities that show up in NT- and in vitro-produced animals.[70,71]These abnormal phenotypes appear to not be transmitted to offspring. The other phenotypes in NT-derived animals appear to be species-specific. Our first transgenic pig was created by oocyte transduction, in vitro fertilization, and culture to the blastocyst stage. She had contracted tendons (this was the first pig in the literature resulting from oocyte maturation, fertilization, and development to the blastocyst all in vitro prior to embryo transfer).[72] This female pig subsequently had 24 offspring, none of which had a contracted tendon (unpublished results); and when this female was cloned, only one of her four clones had a contracted tendon.[42] Other examples of LOS that are not transmitted to offspring include large birth weights in cattle,[73] and obesity in cloned mice.[74] Even when an animal with an abnormal phenotype is cloned, such abnormalities generally do not appear in the resulting offspring.[75] One theory that would describe the mechanism(s) of LOS is aberrant DNA methylation.

Methylation of DNA affects the 3-dimensional structure of the DNA such that transcription is altered, and can affect phenotypic characteristics such as coat color.[76,77] Apparently aberrant DNA methylation is the cause of some of the LOS.[78] We have occasionally observed macroglossia (enlarged tongue) in our cloned pigs. Macroglossia is consistent with Beckwith-Wiedemann syndrome and aberrant IGF2 gene methylation and expression in humans.[79] During normal embryo development, active DNA demethylation occurs in the paternal genome in the zygote followed by gene-specific passive (or dilution) demethylation during early cleavage, and de novo methylation of the inner cell mass cells of the blastocyst.[80] Some cloned embryos fail to recapitulate the normal pattern of global demethylation and gene-specific methylation observed during normal embryogenesis[81,82] and thus result in animals with LOS symptoms. Presumably, these phenotypes are not transmitted to the next generation because the DNA methylation pattern is reestablished during gametogenesis[83,84] or altered during culture of the donor cells or embryo[75] such that a normal appearing animal will result. A more complete description of changes in DNA methylation during normal and NT embryo development will aid in understanding the causes, and hopefully, the prevention of LOS.

Transgenic Pigs

The standby method for making transgenic pigs has been pronuclear injection. Other methods that can be used include sperm mediated gene transfer, oocyte transduction, and the cloning technology. Gene additions are the easiest to perform and do not require nuclear transfer. However, the only method available for making a knockout or an addition at a precise location is that of nuclear transfer. The types of genetic modifications that have been performed in pigs include both gene additions and knockouts. Transgenic pigs have been made (Table 1) for xenotransplantation, transplantation, the production of pharmaceuticals, models of disease, and the improvement of production or quality of meat. The techniques for making these pigs have been pronuclear injection, oocyte transduction, nuclear transfer, and sperm mediated gene transfer. Below I will highlight the genetic modifications that have been constructed, or are being constructed, in our laboratory.

Our first genetic modification was to add the enhanced green fluorescent protein and we used both oocyte transduction[72] and nuclear transfer[15] (Fig. 1). These pigs have proven very valuable for some basic scientific studies that require cells that are already marked so that they can be distinguished from other cells. One example is our or a collaboration with Drs. Michael Young and Henry Klassen. They visit our laboratory in Missouri and collect eyes from fetal transgenic pigs. They then isolate and cultivate retinal progenitor cells. These transgenic retinal progenitor cells are then transferred into the damaged eyes of pigs, and their ability to integrate into and repair that damage by differentiating into rod and cone cells can be easily monitored because the cells are marked.[85,86] Recently other colors such as blue and red have been added to the palette of pig cells that will be useful to mark cells.[87]

The next project was our collaboration with Dr. William Beschorner at Ximerex in Omaha, NB. One goal was to create transgenic pigs that have a suicide gene that is liver specific. Thus we made pigs with liver specific promoters driving either thymidine kinase or cytosine deaminase[88,89] (Fig. 2). Human hepatocytes might then be infused into preimmune fetal transgenic pigs. The human hepatocytes would integrate into and form a chimeric liver. When the animals become adults, they could be treated with the appropriate precursors and kill off the pig hepatocytes, leaving human hepatocytes that might be transferred to humans.

The first knockout pigs were produced by our laboratory in collaboration with Immerge Biotherapeutics.[75,90] The gene that encodes alpha-1,3-galactosyltransferase was knocked out as the gene product places a galactose alpha-1,3-galactose sugar linkage on the surface of the cell. It is this galactose alpha-1,3-galactose sugar linkage that results in hyperacute rejection of pig organs when they are transferred to humans. By knocking out both copies of this gene it was possible to extend the life of the xenograft when the pig heart and kidneys were transferred to baboons.[91,92]

The knockout technology has also been applied to pigs that have the human decay accelerating factor (hDAF) gene addition. These pigs have two copies of hDAF (base genetics are

Table 1. Genetic modifications to swine

Gene	Type*	Function	Use
CD46: Membrane cofactor protein (MCP)[97,98]	A	Complement inhibitor	Xenotransplantation
CD55: Decay accelerating factor (DAF)[99]	A	Complement inhibitor	Xenotransplantation
CD59[100]	A	Complement inhibitor	Xenotransplantation
1,4-N-acetylglucos aminyl-transferase III[101]	A	Reduce alpha-Gal epitopes from forming	Xenotransplantation
Alpha-1,3-galactosyltransferase[90,102]	KO	Prevent alpha-Gal epitopes from forming	Xenotransplantation
Alpha-1,2-fucosyltransferase[103]	A	Reduce alpha-Gal epitopes from forming	Xenotransplantation
Human leukocyte antigen II, DP/DQ[104-106]	A	Prevent xenograft rejection	Xenotransplantation
Enhanced fluorescent proteins[15,42,72,107,108]	A	Tracking of cells	Many basic science uses
Thymidine kinase[89]	A	Liver-specific	Hepatocyte growth and transplantation
Cytosine deaminase[88]	A	Liver-specific	Hepatocyte growth and transplantation
Endothelial cell nitric oxide synthase (eNOS)[109]	A	Overexpress eNOS	Cardiovascular studies
Endothelial cell nitric oxide synthase (eNOS)[110]	KO	Eliminate eNOS	Cardiovascular studies
Human alpha and beta hemoglobin[111]	A	Human hemoglobin production	Pharmaceutical
Protein C[112]	A	Human protein C	Pharmaceutical
Coagulation factor VIII[113]	A	Treat hemophiliacs	Pharmaceutical
Coagulation factor IX[114]	A	Treat hemophiliacs	Pharmaceutical
Rhodopsin[115]	A	Produce defective rhodopsin	Model of retinitis pigmentosa
Cystic fibrosis transmembrane receptor[110]	KO	Produce defective CFTR	Model of cystic fibrosis
Heat shock proten 70.2[116]	A	Protect against heat stress	Production agriculture
Growth hormone[117-120]	A	Improve growth and carcass quality	Meat production
Insulin like growth factor I (IGFI)[121]	A	Improve growth and carcass quality	Meat production
Alpha-lactalbumin[122]	A	Improve milk quality	Increase weaning weight of piglet
Phytase[123]	A	Improve phosphorous utilization	Decrease pollution
Huntington[124]	A	Expand a CAG trinucleotide repeat	Model of Huntington disease
Delta 12 fatty acid desaturase[125]	A	Plant gene expression in a pig	Meat quality

*A: gene addition; KO: gene knockout

Figure 1A. Piglet derived by nuclear transfer that expressed the eGFP gene.[107] Note the yellow color of the nose and hooves (a color version of this image is available online at www.eurekah.com).

Figure 2. Pig that is transgenic for an albumin promoter driving thymidine kinase.[89]

Imutran line 12) and we and our colleagues at Immerge Biotherapeutics have knocked out one copy of alpha-1,3-galactosyltransferase and then cloned the animals. The two females that we have, DAFney and Olga, were born June 28, 2004 and July 6, 2004, respectively (Lai et al, unpublished). These two animals have been donated to the National Swine Resource and Research Center (http://nsrrc.missouri.edu/), and their descendents will be available for distribution to the scientific community.

In collaboration with Drs. Harold Laughlin and Ed Rucker here at MU we have created Yucatan miniature pigs that overexpress endothelial cell nitric oxide synthase (unpublished). These pigs will be very valuable for understanding cardiovascular function in pigs on high fat diets.[93]

Other genetic modifications that we have planned, or are in the process of creating, include: knocking out CFTR to create a model of cystic fibrosis, and knocking out eNOS for cardiovascular studies.

With the ability to add genes, knockdown genes with siRNA, and knockout genes in pigs we are now only limited by our imagination. The above described techniques will be further enhanced with future development of stem cells from pigs,[48] development of the technology to facilitate transplantation of spermatogonial stem cells,[94] artificial pig chromosomes,[95] and the development of cell free systems for reprogramming nuclei.[96] Future genetic modifications will have an even greater impact on medicine and agriculture.

Acknowledgements

Support from the NIH (RR13438, RR18877, RR18276, HL51670 via the University of Iowa) the USDA (2004-05514) and Food for the 21st Century is greatly appreciated.

References

1. Spemann H. Embryonic Development and Induction. New York: Hafner, 1938:210-211.
2. Briggs R, King TJ. Transplantation of living cell nuclei from blastula cells into enucleated frogs' eggs. Proc Natl Acad Sci USA 1952; 38:455-463.
3. Gurdon JB. Nuclear transplantation in eggs and oocytes. J Cell Sci 1986; (Suppl 4):287-318.
4. Prather RS, First NL. Cloning embryos by nuclear transfer. J Reprod Fertility 1990; (Suppl 41):125-134.
5. First NL, Prather RS. Genomic potential in mammals. Differentiation 1991; 48:1-8.
6. Illmensee K, Hoppe PC. Nuclear transplantation in Mus musculus: Developmental potential of nuclei from preimplantation embryos. Cell 1981; 23:9-18.
7. McGrath J, Solter D. Inability of mouse blastomere nuclei transferred to enucleated zygotes to support development in vitro. Science 1983; 226:1317-1319.
8. Willadsen SM. Nuclear transplantation in sheep embryos. Nature 1986; 31:956-962.
9. Prather RS, Barnes FL, Sims ML et al. Nuclear transfer in the bovine embryo: Assessment of donor nuclei and recipient oocyte. Biol Reprod 1987; 37:859-866.
10. Stice SL, Robl JM. Nuclear reprogramming in nuclear transplant rabbit embryos. Biol Reprod 1988; 39:657-664.
11. Prather RS, Sims MM, First NL. Nuclear transplantation in early pig embryos. Biol Reprod 1989; 41:414-418.
12. Sims M, First NL. Production of calves by transfer of nuclei from cultured inner cell mass cells. Proc Natl Acad Sci USA 1994; 91(13):6143-6147.
13. Campbell KHS, McWhir J, Ritchie WA et al. Sheep cloned by nuclear transfer from a cultured cell line. Nature 1996; 380(6569):64-66.
14. Polejaeva IA, Chen SH, Vaught TD et al. Cloned pigs produced by nuclear transfer from adult somatic cells. Nature 2000; 407(6800):86-90.
15. Park KW, Cheong HT, Lai LX et al. Production of nuclear transfer-derived swine that express the enhanced green fluorescent protein. Anim Biotechnol 2001; 12(2):173-181.
16. Prather RS, Sims MM, First NL. Nuclear transplantation in the pig embryo: Nuclear swelling. J Exp Zool 1990; 255:355-358.
17. Prather RS, Sims MM, Maul GG et al. Nuclear lamin antigens are developmentally regulated during porcine and bovine embryogenesis. Biol Reprod 1989; 41:123-132.

18. Prather RS, Rickords LF. Developmental regulation of a snRNP core protein epitope during pig embryogenesis and after nuclear transfer for cloning. Mol Reprod Dev 1992; 33:119-123.
19. Mayes MA, Stogsdill PL, Parry TW et al. Reprogramming of nucleoli after nuclear transfer of pig blastomeres into enucleated oocytes. Dev Biol 1994; 163:542.
20. Parry TW, Prather RS. Carry-over of mRNA during nuclear transfer in pigs. Reproduction, Nutrition, Development 1995; 35(3):313-318.
21. Liu ZH, Hao YH, Lai LX et al. Morphology and dynamics of alpha-tubulin and nuclear mitotic apparatus protein during the first cell cycle in porcine nuclear transfer embryos, parthenogenetic embryos and in vitro fertilization embryos. Biol Reprod 2004; (Special Issue):215.
22. Teranishi T, Tanaka M, Kimoto S et al. Rapid replacement of somatic linker histones with the oocyte-specific linker histone H1 foo in nuclear transfer. Dev Biol 2004; 266(1):76-86.
23. Gao SR, Chung YG, Parseghian MH et al. Rapid H1 linker histone transitions following fertilization or somatic cell nuclear transfer: Evidence for a uniform developmental program in mice. Dev Biol 2004; 266(1):62-75.
24. Bordignon V, Clarke HJ, Smith LC. Factors controlling the loss of immunoreactive somatic histone H1 from blastomere nuclei in oocyte cytoplasm: A potential marker of nuclear reprogramming. Dev Biol 2001; 233(1):192-203.
25. Moore SC, Jason L, Ausio J. The elusive structural role of ubiquitinated histones [Review]. Biochemistry and Cell Biology-Biochimie and Biologie Cellulaire 2002; 80(3):311-319.
26. Baarends WM, Hoogerbrugge TW, Roest HP et al. Histone ubiquitination and chromatin remodeling in mouse spermatogenesis. Dev Biol 1999; 207(2):322-333.
27. Chen HY, Sun JM, Zhang Y et al. Ubiquitination of histone H3 in elongating spermatids of rat testes. J Biol Chem 1998; 273(21):13165-13169.
28. Tovich PR, Oko RJ. Somatic histones are components of the perinuclear theca in bovine spermatozoa. J Biol Chem 2003; 278(34):32431-32438.
29. Prather RS, Sutovsky P, Green JA. Nuclear remodeling and reprogramming in transgenic pig production. Proc Soc Exp Biol Med 2004; 229:1120-1126.
30. Shiels PG, Kind AJ, Campbell KHS et al. Analysis of telomere lengths in cloned sheep. Nature 1999; 399(6734):316-317.
31. Tian XC, Xu J, Yang XZ. Normal telomere lengths found in cloned cattle. Nat Genet 2000; 26(3):272-273.
32. Miyashita N, Shiga K, Fujita T et al. Normal telomere lengths of spermatozoa in somatic cell-cloned bulls. Theriogenology 2003; 59(7):1557-1565.
33. Lanza RP, Cibelli JB, Blackwell C et al. Extension of cell life-span and telomere length in animals cloned from senescent somatic cells. Science 2000; 288(5466):665-669.
34. Miyashita N, Shiga K, Yonai M et al. Remarkable differences in telomere lengths among cloned cattle derived from different cell types. Biol Reprod 2002; 66(6):1649-1655.
35. Le JA, Carter DB, Xu J et al. Telomere lengths in cloned transgenic pigs. Biol Reprod 2004; 70:1589-1593.
36. Kang YK. Typical demethylation events in cloned pig embryos. Clues on species-specific differences in epigenetic reprogramming of a cloned donor genome. J Biol Chem 2001; 276(43):39980-39984.
37. Kang YK, Yeo S, Kim SH et al. Precise recapitulation of methylation change in early cloned embryos. Mol Reprod Dev 2003; 66(1):32-37.
38. Kang YK, Koo DB, Park JS et al. Aberrant methylation of donor genome in cloned bovine embryos. Nat Genet 2001; 28(2):173-177.
39. Kang YK. Influence of oocyte nuclei on demethylation of donor genome in cloned bovine embryos. FEBS Letters 2001; 499(1-2):55-58.
40. Winger QA, Hill JR, Shin TY et al. Genetic reprogramming of lactate dehydrogenase, citrate synthase, and phosphofructokinase mRNA in bovine nuclear transfer embryos produced using bovine fibroblast cell nuclei. Mol Reprod Dev 2000; 56(4):458-464.
41. Park KW, Kuhholzer B, Lai LX et al. Development and expression of the green fluorescent protein in porcine embryos derived from nuclear transfer of transgenic granulosa-derived cells. Anim Reprod Sci 2001; 68(1-2):111-120.
42. Park KW, Lai LX, Cheong HT et al. Mosaic gene expression in nuclear transfer-derived embryos and the production of cloned transgenic pigs from ear-derived fibroblasts. Biol Reprod 2002; 66(4):1001-1005.
43. DeSousa PA, Winger Q, Hill JR et al. Reprogramming of fibroblast nuclei after transfer into bovine oocytes. Cloning 1999; 1:63-69.

44. Daniels R, Hall V, Trounson AO. Analysis of gene transcription in bovine nuclear transfer embryos reconstructed with granulosa cell nuclei. Biol Reprod 2000; 63(4):1034-1040.
45. Daniels R, Hall VJ, French AJ et al. Comparison of gene transcription in cloned bovine embryos produced by different nuclear transfer techniques. Mol Reprod Dev 2001; 60(3):281-288.
46. Humphreys D, Eggan K, Akutsu H et al. Abnormal gene expression in cloned mice derived from embryonic stem cell and cumulus cell nuclei. Proc Natl Acad Sci USA 2002; 99(20):12889-12894.
47. Lee SH, Kim DY, Nam DH et al. Role of messenger RNA expression of platelet-activating factor and its receptor in porcine in vitro-fertilized and cloned embryo development. Biol Reprod 2004.
48. Zhu H, Craig JA, Dyce PW et al. Embryos derived from porcine skin-derived stem cells exhibit enhanced preimplantation development. Biol Reprod 2004; 71(6):1890-7.
49. Hao YH, Lai LX, Mao JD et al. Apoptosis and in vitro development of preimplantation porcine embryos derived in vitro or by nuclear transfer. Biol Reprod 2003; 69(2):501-507.
50. Eggan E, Baldwin K, Tackett M et al. Mice cloned from olfactory sensory neurons. Nature 2004; 428(6978):44-49.
51. Li JS, Ishii T, Feinstein P et al. Odorant receptor gene choice is reset by nuclear transfer from mouse olfactory sensory neurons. Nature 2004; 428(6981):393-399.
52. Hochedlinger K, Jaenisch R. Monoclonal mice generated by nuclear transfer from mature B and T donor cells. Nature 2002; 415(6875):1035-1038.
53. Hochedlinger K, Blelloch R, Brennan C et al. Reprogramming of a melanoma genome by nuclear transplantation. Genes Dev 2004; 18(15):1875-1885.
54. Li L, Connelly MC, Wetmore C et al. Mouse embryos cloned from brain tumors. Cancer Res 2003; 63(11):2733-2736.
55. Whitworth K, Springer GK, Forrester LJ et al. Developmental expression of 2,489 genes during pig embryogenesis: An EST project. Biol Reprod 2004; 71:1230-1243.
56. Whitworth KM, Agca C, Kim JG et al. Transcriptional profiling of pig embryogenesis by using a 15k member unigene set specific for pig reproductive tissues and embryos. and embryos. Biol Reprod 2005; 72(6):1437-51.
57. Krause WJ, Charlson EJ, Sherman DM et al. Three-dimensional reconstruction of mitochondrial aggregates in porcine oocytes, zygotes and early embryos using a personal computer. Zoologischer Anzeiger 1992; 229:21-36.
58. Hyttel P, Niemann H. Ultrastructure of pocrine embryos following development in vitro versus in vivo. Mol Reprod Dev 1990; 27:136-144.
59. Jolliff WJ, Prather RS. Parthenogenic development of in vitro-matured, in vivo-cultured porcine oocytes beyond blastocyst. Biol Reprod 1997; 56(2):544-548.
60. Smith LC, Alcivar AA. Cytoplasmic inheritance and its effects on development and performance. J Reprod Fertility 1993; (Suppl 48):31-43.
61. Meirelles FV, Smith LC. Mitochondrial genotype segregation in a mouse heteroplasmic lineage produced by embryonic karyoplast transplantation. Genetics 1997; 145(2):445-451.
62. Meirelles FV, Smith LC. Mitochondrial genotype segregation during preimplantation development in mouse heteroplasmic embryos. Genetics 1998; 148(2):877-883.
63. Smith LC, Bordignon V, Garcia JM et al. Mitochondrial genotype segregation and effects during mammalian development: Applications to biotechnology. Theriogenology 2000; 53(1):35-46.
64. Meirelles FV, Bordignon V, Watanabe Y et al. Complete replacement of the mitochondrial genotype in a Bos indicus calf reconstructed by nuclear transfer to a Bos taurus oocyte. Genetics 2001; 158(1):351-356.
65. Schatten G. The centrosome and its mode of inheritance- the reduction of the centrosome during gametogenesis and its restoration during fertilization. Dev Biol 1994; 165(2):299-335.
66. Mananadhar G, Schatten H, Lai L et al. Centrosomal protein centrin is not detectable during early cleavages but reappears during late blastocyst stage in porcine embryos. Biol Reprod 2004; (Special Issue):146.
67. Sutovsky P, Manandhar G, Laurincik J et al. Expression and proteasomal degradation of the Major Vault Protein (MVP) in mammalian oocytes and zygotes. Reproduction 2005; 129(3):269-82.
68. Walker SK, Hartwich KM, Seamark RF. The production of unusually large offspring following embryo manipulation - Concepts and challenges. Theriogenology 1996; 45(1):111-120.
69. Wilson JM, Williams JD, Bondioli KR et al. Comparison of birth weight and growth characteristics of bovine calves produced by nuclear transfer (Cloning), embryo transfer and natural mating. Anim Reprod Sci 1995; 38(1-2):73-83.
70. Carter DB, Lai L, Park KW et al. Phenotyping of transgenic cloned pigs. Cloning and Stem Cells 2002; 4:131-145.

71. Carroll JA, Carter DB, Korte SW et al. Evaluation of the acute phase response in cloned pigs following a lipopolysaccharide challenge. Domest Anim Endocrinol 2005; 29(3):564-72

72. Cabot RA, Kuhholzer B, Chan AWS et al. Transgenic pigs produced using in vitro matured oocytes infected with a retroviral vector. Anim Biotechnol 2001; 12(2):205-214.

73. Conway KL. Birth weight of bovine calves produced by nuclear transfer (cloning) and their offspring (embryo transfer). Dissertation Abstracts International 1996; 57-06(B):3462.

74. Tamashiro KLK, Wakayama T, Akutsu H et al. Cloned mice have an obese phenotype not transmitted to their offspring. Nature Medicine 2002; 8(3):262-267.

75. Kolber-Simonds D, Lai L, Watt SR et al. Alpha-1,3-galactosyltransferase null pigs via nuclear transfer with fibroblasts bearing loss of heterozygosity mutations. Proc Natl Acad Sci USA 2004; 101:7335-7340.

76. Gaudet F, Rideout WM, Meissner A et al. Dnmt1 expression in pre- and postimplantation embryogenesis and the maintenance of IAP silencing. Molecular and Cellular Biology 2004; 24(4):1640-1648.

77. Wolff GL, Kodell RL, Moore SR et al. Maternal epigenetics and methyl supplements affect agouti gene expression in a(Vy)/a mice. FASEB Journal 1998; 12(11):949-957.

78. Jaenisch R, Bird A. Epigenetic regulation of gene expression: How the genome integrates intrinsic and environmental signals [Review]. Nat Genet 2003; 33(Suppl S):245-254.

79. Pray LA. Epigenetics: Genome, meet your environment. Scientist 2004; 18(13):14.

80. Santos F, Hendrich B, Reik W et al. Dynamic reprogramming of DNA methylation in the early mouse embryo. Dev Biol 2002; 241(1):172-182.

81. Bourc'his D, Le Bourhis D, Patin D et al. Delayed and incomplete reprogramming of chromosome methylation patterns in bovine cloned embryos. Current Biology 2001; 11(19):1542-1546.

82. Dean W, Santos F, Stojkovic M et al. Conservation of methylation reprogramming in mammalian development: Aberrant reprogramming in cloned embryos. Proc Natl Acad Sci USA 2001; 98(24):13734-13738.

83. Humpherys D, Eggan K, Akutsu H et al. Epigenetic instability in ES cells and cloned mice. Science 2001; 293(5527):95-97.

84. Rideout WM, Eggan K, Jaenisch R. Nuclear cloning and epigenetic reprogramming of the genome. Science 2001; 293(5532):1093-1098.

85. Klassen HJ, Warfving K, Kiilgaard JF et al. Transplantation of retinal progenitor cells from GFP-transgenic pigs to the injured retina of allogeneic recipients. Investigative Ophthalmology and Visual Science 2004; 45(5400 Suppl 2):U646.

86. Shatos MA, Klassen HJ, Schwartz PH et al. Isolation of projenitor cells from retina and brain of the GFP-transgenic pig. Investigative Ophthalmology and Visual Science 2004; 45(5406 Suppl 2):U647.

87. Webster NL, Forni M, Bacci ML et al. Multi-transgenic pigs expressing three fluorescent proteins produced with high efficiency by sperm mediated gene transfer. Mol Reprod Dev 2005; 72:68-72.

88. Beschorner W, Joshi SS, Prather R et al. Selective and conditional depletion of pig cells with transgenic pigs and specific liposomes. Xenotransplantation 2003; 10(5):497.

89. Beschorner W, Prather R, Sosa C et al. Transgenic pigs expressing the suicide gene thymidine kinase in the liver. Xenotransplantation 2003; 10(5):530.

90. Lai LX, Kolber-Simonds D, Park KW et al. Production of alpha-1,3-galactosyltransferase knockout pigs by nuclear transfer cloning. Science 2002; 295(5557):1089-1092.

91. Kuwaki K, Tseng YL, Dor F et al. Heart transplantation in baboons using 1,3-galactosyl transferase gene-knockout pigs as donors: Initial experience. Nature Medicine 2005; 11(1):29-31.

92. Yamada K, Yazawa K, Shimizu A et al. Marked prolongation of porcine renal xenograft survival in baboons through the use of alpha 1,3-galactosyltransferase gene-knockout donors and the cotransplantation of vascularized thymic tissue. Nature Medicine 2005; 11(1):32-34.

93. Henderson KK, Turk JR, Rush JWE et al. Endothelial function in coronary arterioles from pigs with early-stage coronary disease induced by high-fat, high-cholesterol diet: Effect of exercise. Journal of Applied Physiology 2004; 97(3):1159-1168.

94. Dobrinski I, Avarbock MR, Brinster RL. Germ cell transplantation from large domestic animals into mouse testes. Mol Reprod Dev 2000; 57(3):270-279.

95. Poggiali P, Scoarughi GL, Lavitrano M et al. Construction of a swine artificial chromosome: A novel vector for transgenesis in the pig. Biochimie 2002; 84(11):1143-1150.

96. Collas P. Nuclear reprogramming in cell-free extracts. Philosophical Transactions of the Royal Society of London - Series B: Biological Sciences 2003; 358(1436):1389-1395.

97. Schneider-Schaulies J, Martin MJ, Logan JS et al. CD46 transgene expression in pig peripheral blood mononuclear cells does not alter their susceptibility to measles virus or their capacity to downregulate endogenous and transgenic CD46. Journal of General Virology 2000; 81(Part 6):1431-1438.

98. Diamond LE, Quinn CM, Martin MJ et al. A human CD46 transgenic pig model system for the study of discordant xenotransplantation. Transplantation 2001; 71(1):132-142.

99. Langford GA, Yannoutsos N, Cozzi E et al. Production of pigs transgenic for human decay accelerating factor. Transplant Proc 1994; 26(3):1400-1401.

100. Fodor WL, Williams BL, Matis LA et al. Expression of a functional human complement inhibitor in a transgenic pig as a model for the prevention of xenogeneic hyperacute organ rejection. Proc Natl Acad Sci USA 1994; 91(23):11153-11157.

101. Miyagawa S, Murakami H, Murase A et al. Transgenic pigs with human N-acetylglucosaminyltransferase III. Transplant Proc 2001; 33(1-2):742-743.

102. Phelps CJ, Koike C, Vaught TD et al. Production of alpha 1,3-galactosyltransferase-deficient pigs. Science 2003; 299(5605):411-414.

103. Koike C, Kannagi R, Takuma Y et al. Introduction of Alpha(1,2)-fucosyltransferase and its effect on alpha-gal epitopes in transgenic pig. Xenotransplantation 1996; 3(1 Part 2):81-86.

104. Tu CF, Tsuji K, Lee KH et al. Generation of HLA-DP transgenic pigs for the study of xenotransplantation. Inter Surg 1999; 84(2):176-182.

105. Tu CF, Hsieh SL, Lee JM et al. Successful generation of transgenic pigs for human decay-accelerating factor and human leucocyte antigen DQ. Transplant Proc 2000; 32(5):913-915.

106. Lee JM, Tu CF, Yang PW et al. Reduction of human-to-pig cellular response by alteration of porcine MHC with human HLA DPW0401 exogenes. Transplantation 2002; 73(2):193-197.

107. Lai LX, Park KW, Cheong HT et al. Transgenic pig expressing the enhanced green fluorescent protein produced by nuclear transfer using colchicine-treated fibroblasts as donor cells. Mol Reprod Dev 2002; 62(3):300-306.

108. Webster NL, Forni M, Bacci ML et al. Multi-transgenic pigs expressing three fluorescent proteins produced with high efficiency by sperm mediated gene transfer. Mol Reprod Dev 2005; 72(1):68-76.

109. Hao Y, Yong HY, Murphy CN et al. Production of transgenic cloned piglets by using porcine fetal fibroblasts overexpressing endothelial nitric oxide synthase (eNOS). Reprod Fert Dev 2006; 18:109w.

110. Prather RS et al. Work in progress.

111. Sharma A, Martin MJ, Okabe JF et al. An isologous porcine promoter permits high level expression of human hemoglobin in transgenic swine. Bio-Technology 1994; 12(1):55-59.

112. Lee TK, Bangalore N, Velander W et al. Activation of recombinant human protein C. Thrombosis Research 1996; 82(3):225-234.

113. Paleyanda RK, Velander WH, Lee TK et al. Transgenic pigs produce functional human factor Viii in milk. Nat Biotechnol 1997; 15(10):971-975.

114. Lindsay M, Gil GC, Cadiz A et al. Purification of recombinant DNA-derived factor IX produced in transgenic pig milk and fractionation of active and inactive subpopulations. J Chromatogr 2004A; 1026(1-2):149-157.

115. Petters RM, Alexander CA, Wells KD et al. Genetically engineered large animal model for studying cone photoreceptor survival and degeneration in retinitis pigmentosa. Nat Biotechnol 1997; 15(10):965-970.

116. Chen MY, Tu CF, Huang SY et al. Augmentation of thermotolerance in primary skin fibroblasts from a transgenic pig overexpressing the porcine HSP70.2. AJAS (Asian-Australasian Journal of Animal Sciences) 2005; 18(1):107-112.

117. Pursel VG, Wall RJ, Solomon MB et al. Transfer of an ovine metallothionein-ovine growth hormone fusion gene into swine. J Anim Sci 1997; 75(8):2208-2214.

118. Nottle MB, Nagashima H, Verma PJ et al. Production and analysis of transgenic pigs containing a metallothionein porcine growth hormone gene construct. In: Murray JD, Anderson GB, Oberbauer AM et al, eds. Transgenic Animals in Agriculture. Wallingford, Oxon, England: CAB International, 1999:145-156, (ISBN 0-85199-85293-85195).

119. Hirabayashi M, Takahashi R, Ito K et al. A comparative study on the integration of exogenous DNA into mouse, rat, rabbit, and pig genomes. Exp Anim 2001; 50(2):125-131.

120. Cifone D, Dominici FP, Pursel VG et al. Inability of heterologous growth hormone (GH) to regulate GH binding protein in GH-transgenic swine. J Anim Sci 2002; 80(7):1962-1969.

121. Pursel VG, Wall RJ, Mitchell AD et al. Expression of insulin-like growth factor-I in skeletal muscle of transgenic swine. In: Murray JD, Anderson GB, Oberbauer AM et al, eds. Transgenic Animals in Agriculture. 198 Madison Ave, New York, NY: CAB International, 1999:131-144, (10016-4341 USA).

122. Bleck GT. Production of bovine alpha-lactalbumin in the milk of transgenic pigs. J Anim Sci 1998; 76(12):3072-3078.
123. Golovan SP. Pigs expressing salivary phytase produce low-phosphorus manure. Nature Biotechnology 2001; 19(8):741-745, [erratum appears in Nat Biotechnol 2001 Oct;19(10):979].
124. Uchida M, Shimatsu Y, Onoe K et al. Production of transgenic miniature pigs by pronuclear microinjection. Transgenic Res 2001; 10(6):577-582.
125. Saeki K, Matsumoto K, Kinoshita M et al. Functional expression of a Delta 12 fatty acid desaturase gene from spinach in transgenic pigs. Proc Natl Acad Sci USA 2004; 101(17):6361-6366.

CHAPTER 2

Somatic Cell Nuclei in Cloning:
Strangers Traveling in a Foreign Land

Keith E. Latham,* Shaorong Gao and Zhiming Han

Abstract

The recent successes in producing cloned offspring by somatic cell nuclear transfer are nothing short of remarkable. This process requires the somatic cell chromatin to substitute functionally for both the egg and the sperm genomes, and indeed the processing of the transferred nuclei shares aspects in common with processing of both parental genomes in normal fertilized embryos. Recent studies have yielded new information about the degree to which this substitution is accomplished. Overall, it has become evident that multiple aspects of genome processing and function are aberrant, indicating that the somatic cell chromatin only infrequently manages the successful transition to a competent surrogate for gamete genomes. This review focuses on recent results revealing these limitations and how they might be overcome.

Introduction

Cloning of animals by nuclear transfer enjoys a history of more than 50 years, dating back to the pioneering studies of King and Briggs.[1] During these past five decades, some remarkable advances have been realized. Most recently, cloning has been successful in a variety of different mammalian species[2-16] (including sheep, mouse, cow, goat, pig, horse and mule, cat, and dog) and also fish.[17]

Although cloning has been successful in many species, the overall rate of success has remained quite low, in the range of 1-5% for both amphibians and mammals.[18] Many different approaches have been taken to try to improve success, including treatment of donor cells or their nuclei with chemical agents or cellular extracts, employing different methods of depleting the recipient oocyte of its genome, employing different stages of oocyte recipient, employing different cell cycle stages of donor cell, employing different types of donor cells, and employing serial nuclear transfer, including the production of embryonic stem cells, which are then used for a second round of cloning.[19-26] Some of these efforts have met with modest success, but in general they have not yielded a major increase in efficiency. The most notable success has been with the use of serial nuclear transfer and ES cell derivation to achieve cloning with especially refractory donor cell types.[19] The selection of culture medium for cloned embryos has a major impact on success, and the selection of culture system for oocyte maturation and culture may likewise be important.[27-32]

With this database of experience, it is important to consider whether current cloning approaches are destined to be limited to a low rate of success due to consequences of specific procedural steps, and, if so, whether these limitations might be overcome. This review will

*Corresponding Author: Keith E. Latham—The Fels Institute for Cancer Research and Molecular Biology, Department of Biochemistry, 3307 North Broad Street, Philadelphia, Pennsylvania 19140, U.S.A. Email: klatham@temple.edu

Somatic Cell Nuclear Transfer, edited by Peter Sutovsky. ©2007 Landes Bioscience and Springer Science+Business Media.

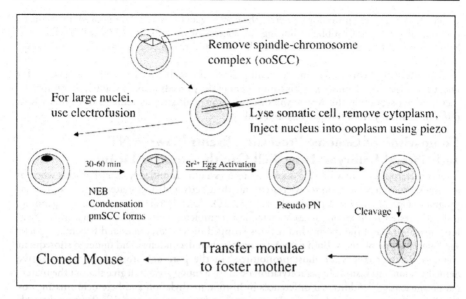

For large nuclei,
use electrofusion

Remove spindle-chromosome
complex (ooSCC)

Lyse somatic cell, remove cytoplasm,
Inject nucleus into ooplasm using piezo

30-60 min

NEB
Condensation
pmSCC forms

Sr²⁺ Egg Activation

Pseudo PN

Cleavage

Cloned Mouse ← Transfer morulae to foster mother ←

Figure 1. Summary of the cloning procedure. The steps shown illustrate the most common approach for mouse. Procedures for other species are similar, but may include differences in oocyte recipient stage, method of depleting oocyte genetic material, method for introducing nuclei, and oocyte activation protocol. Note that the oocyte spindle chromosome complex (ooSCC) is removed, and that a new pseudomeiotic SCC (pmSCC) forms after about one hour.

focus on the basic question of how well the donor genome can interface with the ooplasmic environment to recapitulate the normal series of events that prepare a fertilized embryo for development. The review will also address whether any deficiencies in this recapitulation might be soluble through genetic engineering to create oocytes specifically designed for supporting clone development.

The Cloning Procedure

Cloning procedures in animals today follow a similar series of steps (Fig. 1). The procedure begins with microsurgical removal of the oocyte genetic material, which is then discarded. This can be achieved at different stages, but most often is achieved at the metaphase II stage in mammals. Most often, microsurgical procedures are employed, but chemically assisted "enucleation" (a commonly used misnomer for removing the spindle-chromosome complex; the metaphase II stage oocyte has no nucleus) employing critical concentrations of cytoskeletal inhibitors can also be employed.[33-34] The donor nucleus is then introduced either by electrofusion or by microinjection. Cell fusion is employed commonly in livestock species, and is also employed in the mouse when donor nuclei exceed the size that can be reliably injected without harm to either the nucleus or the oocyte. Microinjection in the mouse is facilitated by use of a piezo pipet driver.[3] In some species, the electrofusion procedure also results in oocyte activation, and additional treatments can be employed to enhance the efficiency of oocyte activation. In the mouse, neither electrofusion nor microinjection lead to oocyte activation, and after nuclear transfer, the nucleus may be allowed to reside in the ooplasm for a period of time. Activation is typically achieved by exposure to strontium chloride for 5-6 hours. Cytoskeletal inhibitors, such as cytochalasin B are employed to suppress extrusion of a polar body, thereby retaining all of the donor cell genetic material in the oocyte. Some investigators have attempted "semi-cloning", wherein polar body extrusion is permitted in an effort to obtain an artificial gamete. However, it appears that chromosome segregation is random and chaotic, and effects of genomic imprinting are neglected, making this approach not feasible.[35-36]

The first step in the procedure, wherein the oocyte genetic material is removed, may constitute one of the biggest problems limiting the success of the procedure. As will be discussed below, essential factors may be depleted at this step, and this may compromise later development.

Another critical component in the cloning procedure is the choice of culture medium for oocytes and for cloned embryos. This component of the procedure likely still requires improvement. A later section in this review will discuss unique culture requirements of clones, which differ dramatically from normal fertilized embryos.

Comparison of Genome Processing Events between NT and Normal Embryos: How Well Can the Stranger Fit In?

At the start of each new life, the ooplasm executes a unique and very special process wherein a totipotent embryonic genome is created and then activated to initiate the developmental program (Fig. 2) (reviewed previously, refs. 37-38). After fertilization, the oocyte genome, initially arrested at the second meiotic metaphase, completes meiosis. The sperm nuclear envelope disappears, and the sperm chromatin is stripped of its protamines and becomes repackaged with histone proteins. Both parental genomes then decondense, and undergo subsequent changes. These changes are most pronounced in the paternal pronucleus, where active demethylation and histone hyperacetylation occur to a much greater degree than in the maternal pronucleus.[39-41] Additional differences in histone methylation and heterochromatin protein 1 content are also seen between the maternal and paternal pronuclei.[39] Progress through the first S phase creates an early capacity for gene transcription, and a low level of transcription is detectable by a variety of methods during the second half of the one-cell stage. A wave of transient gene induction occurs at the early two-cell stage, a time when enhancers are not essential for transcription (reviewed ref. 37). This early wave of transcription may thus represent promiscuous transcription at a time when the embryo lacks a chromatin structure that permits gene regulation. The passage through the second S phase appears to result in the establishment of a regulatable chromatin structure, as enhancers then become necessary for a high rate of transcription (reviewed ref. 37).

Cloning requires that the donor nucleus substitute for both parental genomes in this entire sequence of events. Indeed, the early processing of the donor nucleus encompasses aspects of both maternal and paternal genome processing. Within an hour of nuclear transfer, the nuclear envelope breaks down, reminiscent of the breakdown of the sperm nuclear envelope and the chromatin is repackaged (see below), just as the sperm chromatin is repackaged. However, because the recipient oocyte remains at the metaphase II stage of meiosis, the donor cell chromatin undergoes condensation and directs the formation of a new spindle, thus bringing itself to a state analogous to the metaphase II stage maternal genome. Oocyte activation then leads to the formation of one, two or even more "pseudopronuclei". Although resembling superficially the pronuclei of fertilized embryos, an important distinction is that the pseudopronuclei do not partition separately the maternal and paternal chromosomes. Thus, one would not anticipate any differential processing of the chromatin within the two pseudopronuclei, as occurs with the physically separated maternal and paternal pronuclear chromatin in normal embryos. Given that maternal and paternal pronuclei may normally compete for limited substances in the oocyte,[41] this situation likely results in a qualitatively different mode of genome processing as compared with the fertilized embryo. For example, the chromosomes in the cloned embryo would not be expected to undergo the same degree of DNA demethylation as occurs in the paternal pronucleus, and similarly, the chromosomes would not be expected to undergo a comparable degree of histone hyperacetylation. Accordingly, as development progresses, the cloned embryonic genome would be expected to differ in critical ways from the normal embryonic genome with respect to specific epigenetic modifications. Indeed, such differences have been reported.[42-43] Once pseudopronuclei are formed, the ensuing S phases should provide opportunities for nuclear reprogramming. One key difference is that, because the donor

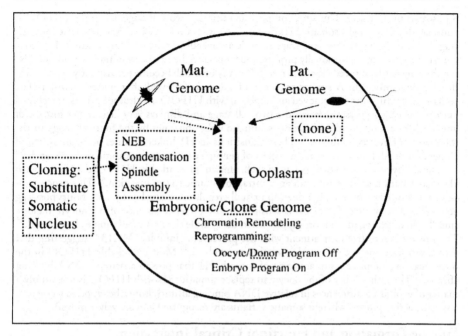

Figure 2. Genome processing in fertilized and cloned embryos. With fertilization (solid arrows), the oocyte must complete its second meiotic division, while the sperm genome becomes decondensed and repackaged by replacement of protamines with histones. Other sperm factors are introduced, including factors that initiate oocyte activation. Genes previously expressed specifically in the oocyte must be silenced. Embryonic genome activation requires chromatin remodeling, which is coupled to the first two rounds of DNA replication. The two pronuclei that form from the oocyte and sperm genomes display different properties (see text). The end result of chromatin remodeling will be a newly formed embryonic genome that can direct the normal embryonic pattern of gene expression. During cloning (dashed arrows), the transferred somatic cell genome undergoes aspects of both maternal and paternal genome processing pathways. The nuclear envelope breaks down (reminiscent of sperm membrane dissolution), the chromosomes condense, and a pseudomeiotic spindle forms, thus recapitulating aspects of oocyte genome processing. No sperm derived factors are involved, and the oocyte is activated artificially. Subsequently, chromatin remodeling occurs, during which process the somatic cell specific gene expression program is theoretically silenced and the embryonic program is established. As described in the text, this reprogramming is incomplete.

genome initially possesses a "mature" chromatin structure, the transiently expressed class of genes may not be induced in clones as they are in fertilized embryos. Whether this is of any consequence to cloned embryo biology is not known.

One additional difference between cloned and fertilized embryos is that the clones will lack any sperm-derived factors. Sperm oocyte activation factors may direct a series of transient calcium rises in the intracellular space,[44] which will thus not occur in cloned constructs. Clones will also lack the sperm-derived centriole.

Early Remodeling of Donor Chromatin: Histone H1 Linker Transitions

The donor genome in cloned constructs recapitulates very well the normal series of H1 linker histone transitions.[45] During normal development, the oocyte-specific H1FOO linker histone is uniquely complexed with the oocyte chromatin from an early stage of oogenesis (day

8 post partum in mice). This situation persists until the two-cell stage, when the H1FOO is eliminated. Interestingly, somatic H1s do not occupy chromatin extensively until the four-cell stage, so that during the two-cell SCC there is an overall paucity of linker histone. This may contribute to the transcriptionally promiscuous state and the first wave of transcription involving the transiently induced class of genes,[46-49] many of which become permanently repressed as development proceeds. A similar sequence of events is followed after nuclear transfer. The somatic linker histones are removed and replaced with H1FOO. As cleavage occurs, the cloned embryo once again passes through the two-cell stage and displays very little linker histone of any kind for a while, and then acquires somatic linker histones by the four-cell stage in the same way as fertilized embryos.[45] These transitions in H1 linker histones occur in 100% of clones examined, indicating that this aspect of genome processing is recapitulated efficiently.[45]

The ability of H1FOO to associate with chromatin is remarkable. H1FOO decorates sperm chromatin within as little as 5 min after intracytoplasmic sperm injection.[45] This process appears to be an active process. It is developmentally regulated, being lost by a few hours post-egg activation. The timing of this loss is sensitive to the presence/absence of the oocyte meiotic spindle, disappearing much more rapidly when the spindle chromosome complex is removed.[45] The process is sensitive to treatment with the proteasome inhibitor MG132, suggesting that active, substrate-specific protein degradation is involved.[50] Most remarkably, H1FOO in the oocyte can even replace a mutant form of somatic H1 that possess a stronger DNA binding affinity.[51] Thus, the ability of the oocyte to replace somatic H1s with H1FOO is not simply a matter of exploiting differences in relative DNA binding affinity, but rather appears to involve a specific active process wherein somatic variants are recognized and actively removed.

Spindle Formation and Function: Critical Interactions between Donor Chromatin and the Ooplasm

While chromatin structure is most often considered in the context of gene regulation, chromatin structure also plays a critical role in maintaining genome integrity. Chromatin-associated proteins interact with the cytoskeleton to direct the formation of meiotic and mitotic spindles, to direct assembly of the chromosomes on the spindle, and to regulate progression through mitosis or meiosis to ensure correct chromosome assembly and segregation.[52-71] In mouse oocytes, numerous microtubule organizing centers (MTOCs) exist that can direct spindle formation in the absence of chromatin.[68] However, when chromatin is present it has the ability to organize these MTOCs, orchestrating and coordinating spindle formation.[68] It is not unreasonable to propose, therefore, that the oocyte chromatin is specialized to organize a meiotic spindle within the oocyte. Thus, removal of the oocyte spindle-chromatin complex (SCC) may deplete the oocyte of essential factors that may be needed for correct spindle assembly.

Studies by Simerly et al[72] indicated that SCC removal might indeed deplete primate oocytes of essential factors. Subsequent studies indicated that some of these factors may not be depleted, but instead that their association with the spindle that forms after nuclear transfer may be defective.[73] It was suggested that this could account for the particularly inefficient nature of primate cloning.[72,73]

More recent studies revealed that SCC removal from mouse oocytes can partially deplete some proteins, but that a majority of those proteins examined are entirely or substantially replenished within a few hours of SCC removal.[74] Interestingly, some proteins fail to associate with the spindle, despite substantial presence revealed by Western blotting. The failure of these proteins to associate with the spindle differed according to whether the donor genome came from a somatic cell or an embryonic cell.[74] The difference in spindle properties is correlated with a difference in the degree of chromosome congression onto the spindle. Additionally, the difference in spindle properties persists at least through the first mitosis, thus creating the potential for aneuploidy in cloned constructs (because polar body extrusion is prevented during oocyte activation, aneuploidy should not arise at that step). Thus, it appears that different donor cells may possess different chromatin structures that differ in their abilities to direct the formation of spindles of normal compositions. A possible role for donor cell centrioles also exists, but because embryonic stem cells and somatic cells display no morphological differences in their centrioles,[75,76] the likelihood of this is not yet clear.

Other effects of SCC removal have been reported. For example, some of the aberrant characteristics of cloned embryos (see below) are ameliorated in tetraploid constructs produced by injecting donor nuclei without removing the oocyte SCC.[27,28,77-79] These include effects on aberrant gene expression and effects on metabolism. These observations suggest that factors other than just DNA and spindle proteins related to chromosome movement are removed when the SCC is removed, and that these factors may participate in regulating a variety of cellular functions in the early embryo.

Overall, the results obtained to date related to spindle formation and function indicate that the somatic donor cell chromatin does not fully mimic oocyte chromatin with respect to the formation of the new spindle before oocyte activation, and neither does it fully mimic the maternal and paternal genomes as the embryo moves into first mitosis. Defects in the formation of the spindle may increase the risk of mitotic errors at each cleavage division, leading to the aneuploidy that has been reported for cloned embryos[80-82] and thus contributing to the overall low success rate of cloning. Additionally, loss of key regulatory factors associated with the SCC may be a factor. Thus, somatic cell nuclei remain strangers in the foreign land of the ooplasm, and seem unable to speak the native language sufficiently well to become fully integrated and assimilated.

Nuclear Reprogramming: How Fast? How Complete?

The oocyte and sperm are two of the most highly differentiated cell types known. Thus, the final task completed by either gamete before entering the final stages of gametogenesis requires a highly specialized repertoire of gene expression. During normal development, the ooplasm is charged with the task of reprogramming the two gamete genomes from these two highly specialized expression states, and establishing a naïve gene expression state from whence to initiate the developmental program. With respect to the maternal genome, chromatin transcriptional silencing and chromatin condensation occur over a protracted period well before metaphase II arrest. With respect to the paternal genome, the chromatin is highly condensed in the sperm and, although it may retain some associated transcription factors, the overall abundance of such factors must be reduced on the sperm chromatin. Additionally, processing of the sperm genome involves a global exchange of chromatin packaging proteins, thus providing the potential for very rapid and very extensive changes in chromatin structure. It is the unique capacity of the oocyte to accomplish this remarkable feat that underlies the success of cloning. However, the beginning substrate for reprogramming, the donor genome, enters the oocyte in a transcriptional state that is very different from that of either gamete. Because the establishment of transcriptionally active and inactive states is a complex process involving hundreds of different transcription and chromatin remodeling factors that vary from one cell type to another, because transcriptional silencing and chromatin remodeling occur in the oocyte long before nuclear transfer, and because the donor genome is unlikely to undergo the same kind of extensive repackaging as the paternal genome, the questions arise as to how well the oocyte is equipped with the necessary abundance and array of factors to deal with all of the various gene expression states within any given type of donor genome, how quickly can this be achieved, and how complete is the end result.

Culture Parameters for Clones

If nuclear reprogramming is highly efficient and highly rapid, then when the cloned embryo attains the stage at which the major transcriptional activation event occurs, the result should be an array of gene expression very similar to that of normal embryos, and consequently very similar phenotypes between cloned and fertilized embryos. Conversely, if reprogramming is slow or incomplete, then the expected result would be an altered pattern of gene expression and a significant divergence in phenotype.

Normal mouse embryos differ substantially from somatic cells with respect to their physiology and their culture requirements. This applies to such basic cellular parameters as osmoregulation, pH regulation, ion transport, carbon substrate utilization, oxygen consumption, and mitochondrial properties,[83-102] all of which has led to the development of highly optimized culture systems that incorporate lower osmolarities, altered ion content, altered carbon substrate compositions, and reduced oxygen concentrations in order to achieve efficient in vitro development while maintaining a normal allocation of cells to the inner cell mass and trophectoderm lineages and a normal pattern of gene expression.[103-106] As a result, significant changes in the phenotypes of cloned embryos should be reflected in their culture requirements.

We[27-29] and others[30] have indeed found that cloned embryos have very different culture requirements, and that these vary according to donor cell type. Clones prepared with cumulus cell nuclei do not develop well in the highly optimized KSOM or CZB media, do slightly better in the older Whitten's medium formulation, and have performed best in the somatic cell culture formulation MEMα.[29] Clones prepared with primary myoblast nuclei do not develop in any of these media, but instead prefer the same somatic cell culture formulation (Ham's F10:DMEM mixture) that is employed to culture the donor cells themselves.[28] A greater degree of success has been achieved in some studies using CZB or KSOM media, but this typically occurs when dimethylsulfoxide (DMSO) is employed as a solvent for applying cytochalasin B during oocyte activation,[27] and thus does not accurately reflect the underlying biology of reprogramming.

One of the more interesting aspects of cloned embryo culture is the unique preference for glucose containing media, even before the first cleavage division. A two-fold greater rate of progression to the two-cell stage is realized when glucose is added to the culture medium.[27] Because fertilized embryos rely almost exclusively on small sugars as substrates for energy metabolism, this requirement for glucose highlights a significant departure of cloned embryos from the normal phenotype.

Along with altered culture requirements, clones display aberrant gene expression indicative of persistent expression of the donor cell phenotype. Cloned embryos prepared with myoblast nuclei express the muscle-specific glucose transporter GLUT4 as early as the one-cell stage, and continue to express this marker through the blastocyst stage.[28] Clones undergo precocious localization of the GLUT1 transporter to the cell surface at the two-cell stage, as opposed to the eight-cell stage seen in normal embryos.[28] Clones also display increased glucose uptake even before the first cleavage division.[28]

Taken together, these results indicate that the phenotype and culture requirements of cloned embryos differ significantly from fertilized embryos and that these differences can arise very early, even before the first cell division. At least some portion of the somatic cell gene expression repertoire fails to be reprogrammed quickly and can be expressed as soon as the embryo becomes competent to undergo gene transcription.

Gene Expression Profile before the First Mitosis

Gene expression during the first cell cycle is almost entirely driven by maternally inherited mRNAs and proteins deposited into the oocyte during oogenesis, with just a few endogenous genes transcribed toward the end of the one-cell stage (reviewed refs. 37,38). A temporally complex change in protein synthesis profiles is evident during the firs cell cycle, and recruitment of maternal mRNAs in an equally temporally complex pattern is likely the reason.[47,49,107] The timing of maternal mRNA recruitment may be coupled to cell cycle progression via the post-translational modifications of mRNA binding proteins and translation factors by cell cycle regulators.[108] One would predict, therefore, that as cloned embryos progress through S phase and prepare for the first mitosis the normal pattern of maternal mRNA recruitment would be observed. The finding that one-cell stage cloned embryos require glucose for efficient cleavage, however, raises the possibility that the one-cell cloned embryo expresses an expanded array of gene products. To evaluate the degree to which cloned embryos display an altered protein synthesis pattern during the one-cell stage, we compared the protein synthesis patterns of cloned embryos, parthenogenetic embryos, and fertilized embryos by L-[^{35}S]methionine labeling at the late one-cell stage and high-resolution quantitative two-dimensional gel electrophoretic analysis using methods similar to those applied in other studies (Fig. 3).[49] We observed a small number of proteins (<5%) that differed between cloned and fertilized embryos (Fig. 3A). Most of these proteins were synthesized at equivalent rates between clones and parthenogenetic controls, with <2% of proteins analyzed differing at the level of two-fold (Fig. 3B). If one accepts a conservative estimate that approximately 9,000 different types of transcripts may exist in the late one-cell,[48] then this would represent alterations in the rates of synthesis of between approximately 161 and 402 proteins. While comparatively small, this number of affected proteins could be sufficient to alter cellular homeostasis and physiology and thus account for the observed alterations in clone embryo phenotype, e.g., culture requirements. The identification of affected post-transcriptional processes (post-translational modifications, stability, mRNA translation) could prove to be an interesting avenue for elucidating controlling mechanisms during normal development, and how such mechanisms may be affected by early gene transcription.

Aberrant Gene Regulation at Later Stages

Aberrant gene expression persists even into later cleavage stages. As mentioned above, some of the aberrantly expressed proteins (e.g., GLUT4) continue to be expressed.[28] Efforts to switch cloned embryos at later stages from somatic cell formulations to embryo culture media have not met with success,[28] indicating that the cloned embryo phenotype remains altered over the long term. Additionally, clones express the somatic form of the DNA methyltransferase DNMT1 at the eight-cell stage.[77] Expression of this protein is normally suppressed post-transcriptionally,[109] indicating that post-transcriptional controls are defective in clones. Studies of imprinted genes[110] and X-linked genes[80] in clones have revealed a failure of a majority of cloned embryos to assume a normal embryonic mode of gene expression before implantation. Many defects in gene expression and functional characteristics are observed in extraembryonic tissues,[111-119] and other alterations in gene expression, imprinting, and DNA methylation are observed at the blastocyst stage and beyond.[110,120-122]

Nuclear-Mitochondrial Cooperativity

As a stranger in a foreign land, the donor genome could make its presence felt by modifying its environment to suit its own needs. One key point where this may occur is with respect to the mitochondrial population. Numerous studies have reported on the fates of donor cell mitochondria, for both intra- and inter-species cloning (review ref. 123). Among the more striking observations is the finding that, not only can donor cell mitochondria persist in the cloned progeny, but in some cases they can actually come to outnumber or even replace the recipient cell mitochondria. For example, with panda → rabbit nuclear transfer, both panda and rabbit mitochondria are detected at the blastocyst stage but by fetal stages (embryos were transferred to feline surrogates) panda mitochondria predominated.[124] In most cases donor mitochondria decrease in abundance as development proceeds. Thus, nuclear-mitochondrial incompatibility could be a limiting factor in cloning, particularly for inter-species combinations. Additionally, mitochondrial heteroplasmy has been suggested as a possible factor limiting cloning success.[125]

Maintenance of Imprinting: Does the Stranger Retain Its Identity?

Numerous studies have revealed aberrant patterns of DNA methylation and imprinted gene regulation in cloned embryos at all stages examined.[110,120-122] These studies collectively reveal that maintenance of DNA methylation patterns may be deficient in clones. Some degree of change in DNA methylation is expected as part of reprogramming gene expression, but it is

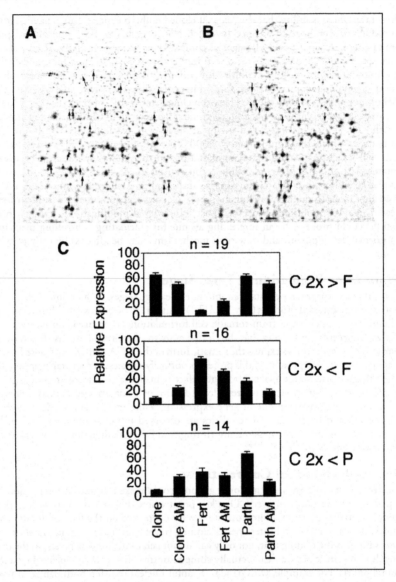

Figure 3. Comparison of protein synthesis patterns of one-cell stage clones, fertilized embryos, and parthenotes. A and B) Two dimensional protein profiles of late 1-cell stage cloned and fertilized embryos, respectively. Proteins that are increased or decreased in cloned embryos relative to fertilized embryos are indicated by upward and downward pointing arrows, respectively. C) Profiles of expression of sets of affected proteins differing between cloned, fertilized, and parthenogenetic embryos. The three graphs portray average expression profiles of the indicated sets; proteins with two-fold or greater elevated rates of synthesis in clones relative to fertilized embryos; proteins with two-fold or greater reduced rates of synthesis in clones relative to fertilized embryos; proteins with two-fold or greater reduced rates of synthesis in clones relative to parthenotes. Average spot intensities for individual spots were normalized to the maximum observed intensity, and then the mean calculated for the entire set. Bars indicate the average and standard error for the set.

essential that parental imprints be retained throughout clone development. Unfortunately, it appears that clones may be deficient in the regulation of DNMT1 import into the nucleus. Eight-cell stage cloned mouse embryos display mosaic nuclear staining for both the somatic and oocyte-specific form of DNMT1, and greatly reduced content of either form for those nuclei that stain positively.[77] This may cause incomplete maintenance of methylation during the fourth round of DNA replication. The long term effect of this would be that during subsequent round of mitosis different blastomeres would segregate different, random combinations of incorrectly and correctly imprinted genes. Those cells with the most normal pattern of imprints may have a selective advantage to contribute to the embryonic lineages, while those with less complete methylation imprinting patterns may become selectively allocated to the extraembryonic lineages. This could account for observed imprinting defects that seem to predominate in cloned placentae,[126] as well as imprinting and growth abnormalities of cloned fetuses and progeny.

Effects of Oocyte Genotype: Different Houses, Different Receptions

The oocyte is clearly a key component in the cloning procedure. It has been known for many years that oocytes of different strains of mice modify both their own genomes and the genomes of incoming sperm in a strain-dependent manner. These effects lead to early differences in gene expression and also can affect DNA methylation (review ref. 127). Surprisingly, comparatively little attention has been given to the effect of oocyte donor genotype on cloning outcome. Several studies have revealed differences in developmental potential of clones prepared with oocytes of different strains, and differences in how such constructs respond to different culture systems.[25,29,128] One notable genetic model of oocyte strain effects is the DDK syndrome, wherein females of the DDK strain display a 95% lethality among embryos sired by nonDDK fathers, but reciprocal crosses are fully viable.[129] Nuclear transfer, ooplasm transfer, and RNA microinjection studies reveal that the DDK oocyte contains an RNA (presumably mRNA) that negatively acts upon the nonDDK paternal genome.[130,131] The result is preimplantation lethality. The DDK syndrome maps to a locus on mouse chromosome 11, which encompasses loci encoding both the maternally expressed (ooplasmic) factor and the target responding locus that is affected in the paternal genome.[132-135] Cloning studies employing oocytes from females that are homozygous for the DDK haplotype on mouse chromosome 11 revealed that the same incompatibility applies to nonDDK somatic cell nuclei as is observed with nonDDK sperm.[136] Thus, the DDK syndrome was recapitulated. However, whereas in the normal DDK syndrome the nonDDK haplotype must be of paternal origin, in the case of cloning, the nonDDK haplotype can be derived from either parent and still exert its effect, indicating loss of the polarity (i.e., imprinting) of the DDK locus either in cumulus cells or in developing clones. Overall, these observations indicate that the oocyte genotype can have a profound effect on the outcome of cloning, and thus that strain-dependent differences likely exist in somatic nucleus modification just as they exist for parental genome modification.

Summary and Perspectives: Making the Strangers Welcome

As summarized in the foregoing pages, cloned embryos are beset with a wide variety of potential problems during the earliest stages of development (Table 1). This list of potential problems continues to grow as different laboratories examine cloned embryos more closely. With such an extensive list of abnormalities, the question arises whether cloning can ever be made more efficient, and whether it can ever be made "safe" for therapeutic applications. We suggest that the answer is likely "yes" on both counts, but that this will require significant progress to be made simultaneously on several fronts. Some of the defects observed could be related to the slow pace of reprogramming, and the subsequent result that cloned embryos are typically being cultured in sub-optimum culture conditions. The limited rates of success of preimplantation and early post-implantation development are likely a consequence of sub-optimum culture. Moreover, imprinting defects may be at least partly soluble by optimizing the culture system. However, clone culture optimization will likely have to be undertaken

Table 1. *Major defects observed in preimplantation stage cloned embryos*

Defects in spindle formation and function
Aneuploidy/Tetraploidy
Slow reprogramming, partial expression of donor cell program
Altered responses to glucose and amino acids in culture
Suboptimum oocyte and embryo culture
Defects in nuclear import
Defects in protein localization
Post-transcriptional gene regulatory defects
Defects in imprinting maintenance
Mitochondrial-nuclear compatibility & heteroplasmy
Oocyte genotype effect

independently for each different donor cell type employed, because the cloned embryo phenotype may vary with donor cell type. The process of culture medium optimization, however, is not trivial. Thus, the question arises whether cloned embryo phenotypes might be altered in order to minimize the need and difficulty of culture medium optimization. One approach to this may be to decipher the molecular basis for the observed defects and then to design genetically engineered oocyte donors that express proteins to correct identified deficiencies. Such oocytes could enhance the success of cloning. Making progress toward this goal should also yield a wealth of new information about basic oocyte and embryo biology.

Acknowledgements

This work was supported by a grant from the National Institutes of Health, National Institute of Child Health and Human Development, HD43092.

References

1. King TJ, Briggs R. Changes in the nuclei of differentiating gastrula cells, as demonstrated by nuclear transplantation. Proc Natl Acad Sci USA 1955; 41:321-325.
2. Wilmut I, Schnieke AE, McWhir J et al. Viable offspring derived from fetal and adult mammalian cells. Nature 1997; 385:810-813.
3. Wakayama T, Perry ACF, Zuccotti M et al. Full-term development of mice from enucleated oocytes injected with cumulus cell nuclei. Nature 1998; 394:369-374.
4. Cibelli JB, Stice SL, Golueke PJ et al. Cloned transgenic calves produced from nonquiescent fetal fibroblasts. Science 1998; 280:1256-1258.
5. Baguisi A, Behboodi E, Melican DT et al. Production of goats by somatic cell nuclear transfer. Nat Biotechnol 1999; 17:456-641.
6. Kato Y, Tani T, Sotomura Y et al. Eight calves cloned from somatic cells of a single adult. Science 1998; 282:2095-2098.
7. Keefer CL, Keyston R, Lazaris A et al. Production of cloned goats after nuclear transfer using adult somatic cells. Biol Reprod 2002; 66:199-203.
8. Kubota C, Yamakuchi H, Todoroki J et al. Six cloned calves produced from adult fibroblast cells after long-term culture. Proc Natl Acad Sci USA 2000; 97:990-995.
9. Park KW, Kuhholzer B, Lai L et al. Development and expression of the green fluorescent protein in porcine embryos derived from nuclear transfer of transgenic granulosa-derived cells. Anim Reprod Sci 2001; 68:111-120.
10. Galli C, Lagutina I, Crotti G et al. Pregnancy: A cloned horse born to its dam twin. Nature 2003; 424:635.
11. Lee JW, Wu SC, Tian XC et al. Production of cloned pigs by whole-cell intracytoplasmic microinjection. Biol Reprod 2003; 69:995-1001.
12. Shin T, Kraemer D, Pryor J et al. A cat cloned by nuclear transplantation. Nature 2002; 415:859.
13. Woods GL, White KL, Vanderwall DK et al. A mule cloned from fetal cells by nuclear transfer. Science 2003; 301:1063.

14. Yin XJ, Tani T, Yonemura I et al. Production of cloned pigs from adult somatic cells by chemically assisted removal of maternal chromosomes. Biol Reprod 2002; 67:442-446.
15. Kasinathan P, Knott JG, Wang Z et al. Production of calves from G1 fibroblasts. Nat Biotechnol 2001; 19:1176-1178.
16. Lee BC, Kim MK, Jang G et al. Dogs cloned from adult somatic cells. Nature 2005; 436:641.
17. Lee KY, Huang H, Ju B et al. Cloned zebrafish by nuclear transfer from long-term-cultured cells. Nat Biotechnol 2002; 20:795-799.
18. Rhind SM, Taylor JE, De Sousa PA et al. Human cloning: Can it be made safe? Nat Rev Genet 2003; 4:855-864.
19. Hochedlinger K, Jaenisch R. Monoclonal mice generated by nuclear transfer from mature B and T donor cells. Nature 2002; 415:1035-1038.
20. Sullivan EJ, Kasinathan S, Kasinathan P et al. Cloned calves from chromatin remodeled in vitro. Biol Reprod 2004; 70:146-153.
21. Enright BP, Sung LY, Chang CC et al. Methylation and acetylation characteristics of cloned bovine embryos from donor cells treated with 5-aza-2'-deoxycytidine. Biol Reprod 2005; 72:944-948.
22. Challah-Jacques M, Chesne P, Renard JP. Production of cloned rabbits by somatic nuclear transfer. Cloning Stem Cells 2003; 5:295-299.
23. Campbell KH, Alberio R. Reprogramming the genome: Role of the cell cycle. Reprod Suppl 2003; 61:477-494.
24. Wakayama T, Yanagimachi R. Effect of cytokinesis inhibitors, DMSO and the timing of oocyte activation on mouse cloning using cumulus cell nuclei. Reproduction 2001; 122:49-60.
25. Wakayama T, Yanagimachi R. Mouse cloning with nucleus donor cells of different age and type. Mol Reprod Dev 2001; 58:376-383.
26. Wakayama T, Yanagimachi R. Cloning the laboratory mouse. Semin Cell Dev Biol 1999; 10:253-258.
27. Chung YG, Mann MR, Bartolomei MS et al. Nuclear-cytoplasmic 'tug-of war' during cloning: Effects of somatic cell nuclei on culture medium preferences in the preimplantation cloned mouse embryo. Biol Reprod 2002; 66:1178-1184.
28. Gao S, Chung YG, Williams JW et al. Somatic cell-like features of cloned mouse embryos prepared with cultured myoblast nuclei. Biol Reprod 2003; 69:48-56.
29. Gao S, Czirr E, Chung YG et al. Genetic variation in oocyte phenotype revealed through parthenogenesis and cloning: Correlation with differences in pronuclear epigenetic modification. Biol Reprod 2004; 70:1162-1170.
30. Heindryckx B, Rybouchkin A, Van Der Elst J et al. Effect of culture media on in vitro development of cloned mouse embryos. Cloning Stem Cells 2001; 3:41-50.
31. Hoshino Y, Uchida M, Shimatsu Y et al. Developmental competence of somatic cell nuclear transfer embryos reconstructed from oocytes matured in vitro with follicle shells in miniature pig. Cloning Stem Cells 2005; 7:17-26.
32. Mastromonaco GF, Semple E, Robert C et al. Different culture media requirements of IVF and nuclear transfer bovine embryos. Reprod Domest Anim 2004; 39:462-467.
33. Gasparrini B, Gao S, Ainslie A et al. Cloned mice derived from embryonic stem cell karyoplasts and activated cytoplasts prepared by induced enucleation. Biol Reprod 2003; 68:1259-1266.
34. Ibanez E, Albertini DF, Overstrom EW. Demecolcine-induced oocyte enucleation for somatic cell cloning: Coordination between cell-cycle egress, kinetics of cortical cytoskeletal interactions, and second polar body extrusion. Biol Reprod 2003; 68:1249-1258.
35. Tateno H, Latham KE, Yanagimachi R. Reproductive semi-cloning respecting biparental origin. A biologically unsound principle. Hum Reprod 2003; 18:472-473.
36. Tateno H, Akutsu H, Kamiguchi Y et al. Inability of mature oocytes to create functional haploid genomes from somatic cell nuclei. Fertil Steril 2003; 79:216-218.
37. Latham KE, Schultz RM. Embryonic genome activation. Front Biosci 2001; 6:D748-759.
38. Latham KE. Mechanisms and control of embryonic genome activation in mammalian embryos. Int Rev Cytol 1999; 193:71-124.
39. Santos F, Peters AH, Otte AP et al. Dynamic chromatin modifications characterise the first cell cycle in mouse embryos. Dev Biol 2005; 280:225-236.
40. Santos F, Hendrich B, Reik W et al. Dynamic reprogramming of DNA methylation in the early mouse embryo. Dev Biol 2002; 241:172-182.
41. Aoki F, Worrad DM, Schultz RM. Regulation of transcriptional activity during the first and second cell cycles in the preimplantation mouse embryo. Dev Biol 1997; 181:296-307.
42. Santos F, Zakhartchenko V, Stojkovic M et al. Epigenetic marking correlates with developmental potential in cloned bovine preimplantation embryos. Curr Biol 2003; 13:1116-1121.

43. Beaujean N, Taylor J, Gardner J et al. Effect of limited DNA methylation reprogramming in the normal sheep embryo on somatic cell nuclear transfer. Biol Reprod 2004; 71:185-193.
44. Ozil JP, Markoulaki S, Toth S et al. Egg activation events are regulated by the duration of a sustained [Ca2+]cyt signal in the mouse. Dev Biol 2005; 282(1):39-54.
45. Gao S, Chung YG, Parseghian MH et al. Rapid H1 linker histone transitions following fertilization or somatic cell nuclear transfer: Evidence for a uniform developmental program in mice. Dev Biol 2004; 266:62-75.
46. Peaston AE, Evsikov AV, Graber JH et al. Retrotransposons regulate host genes in mouse oocytes and preimplantation embryos. Dev Cell 2004; 7:597-606.
47. Evsikov AV, de Vries WN, Peaston AE et al. Systems biology of the 2-cell mouse embryo. Cytogenet Genome Res 2004; 105:240-250.
48. Zeng F, Baldwin DA, Schultz RM. Transcript profiling during preimplantation mouse development. Dev Biol 2004; 272:483-496.
49. Latham KE, Garrels JI, Chang C et al. Quantitative analysis of protein synthesis in mouse embryos. I. Extensive reprogramming at the one- and two-cell stages. Development 1991; 112:921-932.
50. Gao S, Han Z, Kihara M et al. Protease inhibitor MG132 in cloning: No end to the nightmare. Trends Biotechnol 2005; 23:66-68.
51. Becker M, Becker A, Miyara F et al. Differential in vivo binding dynamics of somatic and oocyte-specific linker histones in oocytes and during ES cell nuclear transfer. Mol Biol Cell 2005; 16:3887-3895.
52. Sampath SC, Ohi R, Leismann O. The chromosomal passenger complex is required for chromatin-induced microtubule stabilization and spindle assembly. Cell 2004; 118:187-202.
53. Can A, Semiz O, Cinar O. Centrosome and microtubule dynamics during early stages of meiosis in mouse oocytes. Mol Hum Reprod 2003; 9:749-756.
54. Carazo-Salas RE, Karsenti E. Long-range communication between chromatin and microtubules in Xenopus egg extracts. Curr Biol 2003; 13:1728-1733.
55. Kalitsis P, Fowler KJ, Earle E et al. Partially functional Cenpa-GFP fusion protein causes increased chromosome missegregation and apoptosis during mouse embryogenesis. Chromosome Re 2003; 11:345-357.
56. Askjaer P, Galy V, Hannak E et al. Ran GTPase cycle and importins alpha and beta are essential for spindle formation and nuclear envelope assembly in living Caenorhabditis elegans embryos. Mol Biol Cell 2002; 13:4355-4370.
57. Bilbao-Cortes D, Hetzer M, Langst G et al. Ran binds to chromatin by two distinct mechanisms. Curr Biol 2002; 12:1151-1156.
58. Hetzer M, Gruss OJ, Mattaj IW. The Ran GTPase as a marker of chromosome position in spindle formation and nuclear envelope assembly. Nat Cell Biol 2002; 4:E177-184.
59. Kim BK, Lee YJ, Cui XS et al. Chromatin and microtubule organisation in maturing and preactivated porcine oocytes following intracytoplasmic sperm injection. Zygote 2002; 10:123-129.
60. Combelles CM, Albertini DF. Microtubule patterning during meiotic maturation in mouse oocytes is determined by cell cycle-specific sorting and redistribution of gamma-tubulin. Dev Biol 2001; 239:281-294.
61. Jones MH, He X, Giddings TH et al. Yeast Dam1p has a role at the kinetochore in assembly of the mitotic spindle. Proc Natl Acad Sci USA 2001; 98:13675-13680.
62. Nachury MV, Maresca TJ, Salmon WC et al. Importin beta is a mitotic target of the small GTPase Ran in spindle assembly. Cell 2001; 104:95-106.
63. Gruss OJ, Carazo-Salas RE, Schatz CA et al. Ran induces spindle assembly by reversing the inhibitory effect of importin alpha on TPX2 activity. Cell 2001; 104:83-93.
64. Carazo-Salas RE, Guarguaglini G, Gruss OJ et al. Generation of GTP-bound Ran by RCC1 is required for chromatin-induced mitotic spindle formation. Nature 1999; 400:178-181.
65. Cutts SM, Fowler KJ, Kile BT et al. Defective chromosome segregation, microtubule bundling and nuclear bridging in inner centromere protein gene (Incenp)-disrupted mice. Hum Mol Genet 1999; 8:1145-1155.
66. Williams BC, Murphy TD, Goldberg ML et al. Neocentromere activity of structurally acentric mini-chromosomes in Drosophila. Nat Genet 1998; 18:30-37.
67. Heald R, Tournebize R, Blank T et al. Self-organization of microtubules into bipolar spindles around artificial chromosomes in Xenopus egg extracts. Nature 1996; 382:420-425.
68. Brunet S, Polanski Z, Verlhac MH et al. Bipolar meiotic spindle formation without chromatin. Curr Biol 1998; 8:1231-1234.
69. Woods LM, Hodges CA, Baart E et al. Chromosomal influence on meiotic spindle assembly: Abnormal meiosis I in female Mlh1 mutant mice. J Cell Biol 1999; 145:1395-1406.

70. Yarm FR. Plk phosphorylation regulates the microtubule-stabilizing protein TCTP. Mol Cell Biol 2002; 22:6209-6221.
71. Craig R, Norbury C. The novel murine calmodulin-binding protein Sha1 disrupts mitotic spindle and replication checkpoint functions in fission yeast. J Cell Sci 1998; 111:3609-3619.
72. Simerly C, Dominko T, Navara C et al. Molecular correlates of primate nuclear transfer failures. Science 2003; 300:297.
73. Simerly C, Navara C, Hwan Hyun S et al. Embryogenesis and blastocyst development after somatic cell nuclear transfer in nonhuman primates: Overcoming defects caused by meiotic spindle extraction. Dev Biol 2004; 276:237-252.
74. Miyara F, Han Z, Gao S et al. Nonequivalence of embryonic and somatic cell nuclei affecting spindle composition in clones. Dev Biol 2006; 289:206-217.
75. Baharvand H, Matthaei KI. The ultrastructure of mouse embryonic stem cells. Reprod Biomed Online 2003; 7:330-335.
76. Sathananthan H, Pera M, Trounson A. The fine structure of human embryonic stem cells. Reprod Biomed Online 2002; 4:56-61.
77. Chung YG, Ratnam S, Chaillet JR et al. Abnormal regulation of DNA methyltransferase expression in cloned mouse embryos. Biol Reprod 2003; 69:146-153.
78. Latham KE. Early and delayed aspects of nuclear reprogramming during cloning. Biol Cell 2005; 97:119-132.
79. Kang YK, Koo DB, Park JS et al. Influence of oocyte nuclei on demethylation of donor genome in cloned bovine embryos. FEBS Let 2001; 499:55-58.
80. Nolen LD, Gao S, Han Z et al. X chromosome reactivation and regulation in cloned embryos. Dev Biol 2005; 279:525-540.
81. Shi W, Dirim F, Wolf E et al. Methylation reprogramming and chromosomal aneuploidy in in vivo fertilized and cloned rabbit preimplantation embryos. Biol Reprod 2004; 71:340-347.
82. Booth PJ, Viuff D, Tan S et al. Numerical chromosome errors in day 7 somatic nuclear transfer bovine blastocysts. Biol Reprod 2003; 68:922-928.
83. Brackett BG. In vitro culture of the zygote and embryo. In: Mastroianni Jr L, Biggers JD, eds. Fertilization and embryonic development in vitro. New York: Plenum Press, 1981:63-79.
84. Van Winkle LJ. Amino acid transport regulation and early embryo development. Biol Reprod 2001; 64:1-12.
85. Baltz JM, Biggers JD, Lechene C. A novel H+ permeability dominating intracellular pH in the early mouse embryo. Development 1993; 118:1353-1361.
86. Baltz JM, Biggers JD, Lechene C. Relief from alkaline load in two-cell stage mouse embryos by bicarbonate/chloride exchange. J Biol Chem 1991; 266:17212-17217.
87. Baltz JM, Biggers JD, Lechene C. Two-cell stage mouse embryos appear to lack mechanisms for alleviating intracellular acid loads. J Biol Chem 1991; 266:6052-6057.
88. Edwards LJ, Williams DA, Gardner DK. Intracellular pH of the mouse preimplantation embryo: Amino acids act as buffers of intracellular pH. Hum Reprod 1998; 13:3441-3448.
89. Phillips KP, Baltz JM. Intracellular pH regulation by HCO_3^-/Cl^- exchange is activated during early mouse zygote development. Dev Biol 1999; 208:392-405.
90. Zhao Y, Baltz JM. Bicarbonate/chloride exchange and intracellular pH throughout preimplantation mouse embryo development. Am J Physiol 1996; 271:C1512-1520.
91. Zhao Y, Chauvet PJ, Alper SL et al. Expression and function of bicarbonate/chloride exchangers in the preimplantation mouse embryo. J Biol Chem 1995; 270:24428-24434.
92. Shepard TH, Muffley LA, Smith LT. Mitochondrial ultrastructure in embryos after implantation. Hum Reprod 2000; 15(Suppl 2):218-228.
93. Sathananthan AH, Trounson AO. Mitochondrial morphology during preimplantational human embryogenesis. Hum Reprod 2000; 15(Suppl 2):148-159.
94. Matsumoto H, Shoji N, Sugawara S et al. Microscopic analysis of enzyme activity, mitochondrial distribution and hydrogen peroxide in two-cell rat embryos. J Reprod Fertil 1998; 113:231-238.
95. Shepard TH, Muffley LA, Smith LT. Ultrastructural study of mitochondria and their cristae in embryonic rats and primate (N. nemistrina). Anat Rec 1998; 252:383-392.
96. Hillman N, Tasca RJ. Ultrastructural and autoradiographic studies of mouse cleavage stages. Am J Anat 1969; 126:151-173.
97. Donnay I, Leese HJ. Embryo metabolism during the expansion of the bovine blastocyst. Mol Reprod Dev 1999; 53:171-178.
98. Houghton FD, Thompson JG, Kennedy CJ et al. Oxygen consumption and energy metabolism of the early mouse embryo. Mol Reprod Dev 1996; 44:476-485.
99. Martin KL, Leese HJ. Role of glucose in mouse preimplantation embryo development. Mol Reprod Dev 1995; 40:436-443.

100. Leese HJ, Barton AM. Pyruvate and glucose uptake by mouse ova and preimplantation embryos. J Reprod Fertil 1984; 72:9-13.
101. Lequarre AS, Grisart B, Moreau B et al. Glucose metabolism during bovine preimplantation development: Analysis of gene expression in single oocytes and embryos. Mol Reprod Dev 1997; 48:216-226.
102. Brown JJ, Whittingham DG. The roles of pyruvate, lactate and glucose during preimplantation development of embryos from F1 hybrid mice in vitro. Development 1991; 112:99-105.
103. Lawitts JA, Biggers JD. Optimization of mouse embryo culture media using simplex method. J Reprod Fert 1991; 91:543-546.
104. Ho Y, Wigglesworth K, Eppig JJ et al. Preimplantation development of mouse embryos in KSOM: Augmentation by amino acids and analysis of gene expression. Mol Reprod Dev 1995; 41:232-238.
105. Chatot CL, Ziomek CA, Bavister BD et al. An improved culture medium supports development of random-bred 1-cell mouse embryos in vitro. J Reprod Fertil 1989; 86:679-688.
106. Lane M, Gardner DK, Hasler MJ et al. Use of G1.2/G2.2 media for commercial bovine embryo culture: Equivalent development and pregnancy rates compared to coculture. Theriogenology 2003; 60:407-419.
107. Oh B, Hwang S, McLaughlin J et al. Timely translation during the mouse oocyte-to-embryo transition. Development 2000; 127:3795-3803.
108. Mendez R, Barnard D, Richter JD. Differential mRNA translation and meiotic progression require Cdc2-mediated CPEB destruction. EMBO J 2002; 21:1833-1844.
109. Ratnam S, Mertineit C, Ding F et al. Dynamics of Dnmt1 methyltransferase expression and intracellular localization during oogenesis and preimplantation development. Dev Biol 2002; 245:304-314.
110. Mann MR, Chung YG, Nolen LD et al. Disruption of imprinted gene methylation and expression in cloned preimplantation stage mouse embryos. Biol Reprod 2003; 69:902-914.
111. Stojkovic M, Buttner M, Zakhartchenko V et al. Secretion of interferon-tau by bovine embryos in long-term culture: Comparison of in vivo derived, in vitro produced, nuclear transfer and demi-embryos. Anim Reprod Sci 1999; 55:151-162.
112. Hill JR, Burghardt RC, Jones K et al. Evidence for placental abnormality as the major cause of mortality in first-trimester somatic cell cloned bovine fetuses. Biol Reprod 2000; 63:1787-1794.
113. De Sousa PA, King T, Harkness L et al. Evaluation of gestational deficiencies in cloned sheep fetuses and placentae. Biol Reprod 2001; 65:23-30.
114. Heyman Y, Chavatte-Palmer P, LeBourhis D et al. Frequency and occurrence of late-gestation losses from cattle cloned embryos. Biol Reprod 2002; 66:6-13.
115. Humpherys D, Eggan K, Akutsu H et al. Abnormal gene expression in cloned mice derived from embryonic stem cell and cumulus cell nuclei. Proc Natl Acad Sci USA 2002; 99:12889-12894.
116. Ogura A, Inoue K, Ogonuki N et al. Phenotypic effects of somatic cell cloning in the mouse. Cloning Stem Cells. 2002; 4:397-405.
117. Lee RS, Peterson AJ, Donnison MJ et al. Cloned cattle fetuses with the same nuclear genetics are more variable than contemporary half-siblings resulting from artificial insemination and exhibit fetal and placental growth deregulation even in the first trimester. Biol Reprod 2004; 70:1-11.
118. Ravelich SR, Shelling AN, Ramachandran A et al. Altered placental lactogen and leptin expression in placentomes from bovine nuclear transfer pregnancies. Biol Reprod 2004; 71:1862-1869.
119. Hall VJ, Ruddock NT, French AJ. Expression profiling of genes crucial for placental and preimplantation development in bovine in vivo, in vitro, and nuclear transfer blastocysts. Mol Reprod Dev 2005; 72:16-24.
120. Humpherys D, Eggan K, Akutsu H et al. Epigenetic instability in ES cells and cloned mice. Science 2001; 293:95-97.
121. Rideout IIIrd WM, Eggan K, Jaenisch R. Nuclear cloning and epigenetic reprogramming of the genome. Science 2001; 293:1093-1098.
122. Ohgane J, Wakayama T, Kogo Y et al. DNA methylation variation in cloned mice. Genesis 2001; 30:45-50.
123. Latham KE. Cloning: Questions answered and unsolved. Differentiation 2004; 72:11-22.
124. Chen DY, Wen DC, Zhang YP et al. Interspecies implantation and mitochondria fate of panda-rabbit cloned embryos. Biol Reprod 2002; 67:637-642.
125. Cummins JM. Mitochondria: Potential roles in embryogenesis and nucleocytoplasmic transfer. Hum Reprod Update 2001; 7:217-228.
126. Inoue K, Kohda T, Lee J et al. Faithful expression of imprinted genes in cloned mice. Science 2002; 295:297.
127. Latham KE. Stage-specific and cell type-specific aspects of genomic imprinting effects in mammals. Differentiation 1995; 59:269-282.

128. Bruggerhoff K, Zakhartchenko V, Wenigerkind H et al. Bovine somatic cell nuclear transfer using recipient oocytes recovered by ovum pick-up: Effect of maternal lineage of oocyte donors. Biol Reprod 2002; 66:367-373.

129. Wakasugi N. A genetically determined incompatibility system between spermatozoa and eggs leading to embryonic death in mice. J Reprod Fertil 1974; 41:85-96.

130. Renard JP, Baldacci P, Richoux-Duranthon V et al. A maternal factor affecting mouse blastocyst formation. Development 1994; 120:797-802.

131. Babinet C, Richoux V, Guenet JL et al. The DDK inbred strain as a model for the study of interactions between parental genomes and egg cytoplasm in mouse preimplantation development. Development 1990; (Suppl):81-87.

132. Baldacci PA, Richoux V, Renard JP et al. The locus Om, responsible for the DDK syndrome, maps close to Sigje on mouse chromosome 11. Mamm Genome 1992; 2:100-5.

133. Pardo-Manuel de Villena F, de la Casa-Esperon E, Verner A et al. The maternal DDK syndrome phenotype is determined by modifier genes that are not linked to Om. Mamm Genome 1999; 10:492-497.

134. Pardo-Manuel de Villena F, Naumova AK, Verner AE et al. Confirmation of maternal transmission ratio distortion at Om and direct evidence that the maternal and paternal "DDK syndrome" genes are linked. Mamm Genome 1997; 8:642-646.

135. Pardo-Manuel de Villena F, de La Casa-Esperon E, Williams JW et al. Heritability of the maternal meiotic drive system linked to Om and high-resolution mapping of the Responder locus in mouse. Genetics 2000; 155:283-289.

136. Gao S, Wu G, Han Z et al. Recapitulation of the ovum mutant (Om) phenotype and loss of Om locus polarity in cloned mouse embryos. Biol Reprod 2005; 72:487-491.

CHAPTER 3

Cloning Cattle:
The Methods in the Madness

Björn Oback* and David N. Wells

Abstract

Somatic cell nuclear transfer (SCNT) is much more widely and efficiently practiced in cattle than in any other species, making this arguably the most important mammal cloned to date. While the initial objective behind cattle cloning was commercially driven—in particular to multiply genetically superior animals with desired phenotypic traits and to produce genetically modified animals—researchers have now started to use bovine SCNT as a tool to address diverse questions in developmental and cell biology. In this paper, we review current cattle cloning methodologies and their potential technical or biological pitfalls at any step of the procedure. In doing so, we focus on one methodological parameter, namely donor cell selection. We emphasize the impact of epigenetic and genetic differences between embryonic, germ, and somatic donor cell types on cloning efficiency. Lastly, we discuss adult phenotypes and fitness of cloned cattle and their offspring and illustrate some of the more imminent commercial cattle cloning applications.

The Importance of Cattle Cloning

Within less than a decade, mammalian nuclear transfer (NT) cloning has expanded from a handful of research projects, pursued by a few laboratories in the USA, the UK, and Japan into a world-wide research field on its own, spanning over 160 laboratories across at least 37 nations. Over 75% of these cloning laboratories are working on livestock (cattle, pig, goat, sheep, and buffalo), about 30% on laboratory animals (mouse, rabbit, monkey, and rat) and another 24% on at least 25 other species (Table 1). Since the NT procedures for different species are often very similar, many laboratories are involved in cloning more than one species. Currently, around 80 laboratories across 24 countries are cloning cattle, accounting for almost half of all cloning organizations. This number is 2-3 times higher than for the next most widely cloned species, mouse (around 30 laboratories) and pig (around 25 laboratories). Cattle NT has dominated the number of cloning publications in the last decade, accounting for an annual average of about 30% of PubMed-listed publications in the field (Fig. 1). The notable exception was 1997 (the "Dolly-year") when sheep cloning was overrepresented in the literature. Mouse and pig account for an annual average of 16% and 13% of NT publications, respectively. The total number of publications for each species generally correlates very well with the number of groups practicing NT in the same species (Fig. 2). Occasionally, the number of publications disproportionately exceeds the number of labs involved (e.g., sheep and human), indicating either some highly productive labs or additional publications from groups that do not actually work with the species. Overall, the total number of publications exponentially increased after 1997,

*Corresponding Author: Björn Oback—Reproductive Technologies, AgResearch Ltd.,
Ruakura Research Centre, East Street, Private Bag 3123, Hamilton, New Zealand.
Email: bjorn.oback@agresearch.co.nz

Somatic Cell Nuclear Transfer, edited by Peter Sutovsky. ©2007 Landes Bioscience
and Springer Science+Business Media.

Table 1. List of mammalian species currently being cloned

Common Name	Latin Name	No. Labs[1]	%[2]
Cattle (bovine)	*Bos taurus*	79	49.1
Mouse	*Mus musculus*	28	17.4
Pig	*Sus scrofa*	24	14.9
Rabbit	*Oryctolagus cuniculus*	14	8.7
Goat	*Capra hircus*	9	5.6
Sheep	*Ovis aries*	8	5.0
Rhesus monkey	*Macaca mulatta*	6	3.1
Horse	*Equus caballus*	5	3.1
Rat	*Rattus norwegicus*	5	3.1
Cat	*Felis silvestris catus*	5	3.1
Swamp Buffalo	*Bubalus bubalis*	4	2.5
Cattle (zebu)	*Bos indicus*	4	2.5
Human	*Homo sapiens*	3	1.9
Long-tailed Macaque	*Macaca fascicularis*	3	1.9
Argali Sheep	*Ovis ammon*	2	1.2
Asian Elephant	*Elephus maximus*	1	0.6
Banteng	*Bos javanicus*	1	0.6
Black Bear	*Ursus americanus*	1	0.6
Dog	*Canis lupus familiaris*	1	0.6
Ferret	*Mustela putorius*	1	0.6
Gaur	*Bos gaurus*	1	0.6
Giant Panda	*Ailuropoda melanoleuca*	1	0.6
Hairy-nose Wombat	*Lasiorhinus kreftii*	1	0.6
Llama	*Lama glama*	1	0.6
Mink Whale	*Balaenoptera acuturostrata*	1	0.6
Mouflon	*Ovis orientalis musimon*	1	0.6
Mountain Bongo	*Tragelaphus eurycerus*	1	0.6
Mule	*Equus assinus x Equus caballus*	1	0.6
North Goat	*Capra ibex*	1	0.6
Camel	*Camelus dromedaries*	1	0.6
Red Deer	*Cervus elaphus*	1	0.6
Saola	*Pseudoryx nghetinhensis*	1	0.6
Tasmanian Tiger	*Thylacinus cynocephalus*	1	0.6
White-tail Deer	*Odocoileus virginianus*	1	0.6
Wildcat	*Felis silvestris lybica*	1	0.6
Yak	*Poephagus mutus*	1	0.6

[1] The total number of cloning laboratories in 2005, based on searching PubMed, Web of Science® [v3.0], conference proceedings and other records, was 161; [2] Proportion of the number of groups (out of the total 161) engaged in cloning a particular species.

peaked in 2003, and appears to be slightly declining since then. Based solely on past research investment and output, i.e., the number of labs and their publications, cloned species can be ranked in order of decreasing importance: cattle, mouse, pig, rabbit, and goat/sheep.

The initial objective behind cattle somatic cell NT (SCNT) was mainly commercially driven —namely to multiply genetically superior animals with desired phenotypic traits and to produce genetically modified animals—but the field has now considerably widened its scope. While most cloning research is still published in reproductive biology journals, an increasing number of researchers use SCNT as a method to address diverse questions in developmental

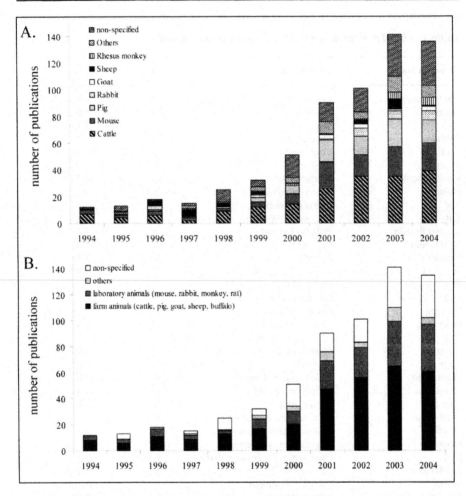

Figure 1. Increasing number of cloning publications over time. A) Number of cloning publications per species as a proportion of the total. B) Number of cloning publications per subject group as a proportion of the total. Only peer-reviewed publications listed in PubMed were included. The category "nonspecified" includes papers that do not address one particular species (e.g., general reviews). "Others" comprises the remaining 28 species listed in Table 1.

and cell biology, such as the nature of epigenetic information, imprinting during gametogenesis, the mechanisms regulating cell plasticity, differentiation and regeneration, and even cancer. We will refer to some examples throughout the text. Naturally, such basic questions are best answered using the laboratory mouse with its short gestation period, large litter size, precisely described embryonic development and well-characterised, easily manipulated genome. However, for all its obvious advantages, there is also one major drawback - mice are very difficult to clone. Although first reported in 1998,[1] still only a handful of laboratories worldwide can reproducibly clone viable mice from somatic cells. Cattle SCNT, on the other hand, is much more routinely practiced and with about 3-5 fold higher average efficiencies. While practice certainly makes perfect, this can not entirely explain the difference in cloning efficiency, especially since some mouse SCNT labs have been practicing for years and still cannot achieve the 10-20% cloning efficiency of many cattle SCNT labs (Table 2, Fig. 3). It is more likely that the

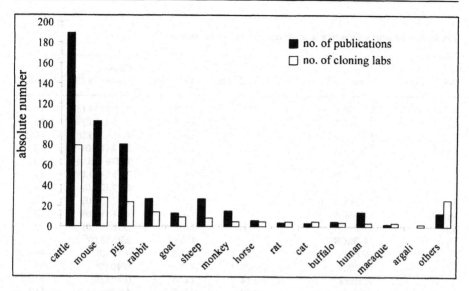

Figure 2. Ranking of cloned species by number of publications and laboratories from 1994-2004. The number of peer-reviewed publications was based on searching PubMed. The number of cloning laboratories was based on searching PubMed, Web of Science® [v3.0], conference proceedings and other records and includes those that may no longer be active in NT cloning. The category "others" comprises another 21 species listed in Table 1.

underlying reasons are based on fundamental species differences in timing of early preimplantation development, for example, the early onset of embryonic genome activation, and rapid preimplantation development in mouse compared to cattle.[2]

Cattle Cloning Methodology

With so many laboratories practicing bovine NT, no two procedures are exactly the same. First of all, there are technical differences resulting from different protocols used. Three types of protocols can be distinguished, mainly based on the method of "enucleation", i.e., removal of the genetic material from the recipient egg to obtain a cytoplast, irrespective of whether that genetic material is surrounded by a nuclear envelope or not: (1) conventional zona-intact NT, (2) zona-free NT, and (3) "hand-made cloning" (HMC). The first method was originally developed for NT into enucleated mouse zygotes[3] and later adopted for enucleated metaphase II (MII)-arrested oocytes, first in sheep,[4] then in cattle.[5] Enucleation is performed by aspirating the maternal chromosomes and surrounding cytoplasm in a small plasma membrane envelope. This method has essentially been used for over 20 years and is still by far the most popular. A simplified zona-free variation of this method is available which effectively doubles the throughput in cloned embryo and cloned offspring production.[6] The zona-free modification offers increased throughput, ease of operation, and reproducibility. It is also easier to learn for beginners with no previous micromanipulation experience. The third method is a more radical deviation from other procedures and relies on manually bisecting the zona-free oocyte with a microblade and discarding the chromatin-containing half. Two enucleated demi-cytoplasts are then fused to reconstitute the original volume before NT. The principle was first exemplified using blastomere donor cells[7] and later adopted and refined for SCNT in cattle.[8]

Different NT protocols confound the search for biological causes underlying low cloning efficiency and a higher degree of protocol standardization across research groups would be desirable. However, even if a standard protocol could be established as ideal cloning practice, there will still be slight variations in its implementation. Differences in experimental nuances

Table 2. *Cloning efficiency and donor cell type in mouse*

Cell Type	Genotype[a]	Cell Line	Cell Cycle	Cloning Efficiency (%)[b]	Ref.
1-cell	F1-B6C3		G_1/S	31/91 (34.0)	15
2-cell	F1-B6C3		G_1/S	5/19 (26.0)	15
	F1-B6CB x ICR		G_1/S	10/34 (29.0)	16
4-cell	F1-B6C3		G_1/S	2/14 (13.0)	15
	F1-B6CB		G_1/S	12/76 (16.0)	185
	F1-B6CB		M	25/58 (43.0)[c]	21
	F1-B6CB x ICR		G_1/S	6/27 (22.0)	16
8-cell	F1-B6CB x ICR		G_1/S	3/17 (18.0)	16
Morula	F1-B6CB		G_1/S	2/21 (10.0)[c]	185
ICM	F1-B6CB		G_1/S	2/18 (11.0)[c]	27
TE	F1-B6CB		G_1/S	2/25 (8.0)[c]	27
ES	*BALB/c*	*v39.7*	*G1/S*	*0/34 (0)*	*36*
	C57BL/6	*v26.2*	*G1/S*	*0/40 (0)*	*36*
	129/ola	*E14*	*G2/M*	*0/135 (0)*	*112*
	129/ola	*E14*	*G1/S*	*1/179 (0.6)*	*112*
	129/ola	*HM-1*	*G1/S*	*5/197 (2.5)*	*111*
	129/Sv	*J1, v18.6*	*G1/S*	*0/153 (0)*	*36,37*
	129/Sv	*NR2, EGFP*	*M*	*2/547 (0.4)*	*38*
	F1-129/Sv x 129/Sv-CP	R1	M	1/47 (2.1)	112
	F1-129/Sv x 129/Sv-CP	R1	M	4/195 (2.0)	40
	F1-129/Sv x 129/Sv-CP	R1	M	0/332 (0)	38
	F1-129/Sv x 129/Sv-CP	R1	G_1/S	12/265 (4.5)	112
	F1-129/Sv x 129/Sv-CP	R1	G_1/S	1/178 (0.6)	111
	F1-129/Sv x BALB/c	v17.2	G_1/S	1/21 (5.0)	36
	F1-129/Sv x C57BL/6	v6.5	G_1/S	7/34 (21.0)	37
	F1-129/Sv x C57BL/6	v6.5	G_1/S	15/79 (19.0)	36
	F1-129/Sv x C57BL/6	v6.5	G_1/S	5/149 (3.4)	186
	F1-129/Sv x C57BL/6	129B6	G_1/S	2/18 (11.0)	36
	F1-129/Sv x FVB	v8.1	G_1/S	2/19 (11.0)	36
	F1-C57BL/6 x BALB/c	v30.30	G_1/S	0/5 (0)	36
	F1-B6C3	Nara 5-12	M	3/1176 (0.3)	38
	F1-B6CB	TT2	M	14/232 (6.0)	39
	F1-B6CB	TT2	M	9/213 (4.2)[c]	39
	F1-B6CB	TT2	M	0/97 (0)	187
	F1-B6CB	TT2	G_1/S	2/169 (1.2)	187
PGC (10.5)	F1-129 x B6D2		G_1	2/441 (0.5)	48
PGC (10.5)	F1-129 x B6D2		M	0/122 (0)	48
PGC (10.5)	F1-B6CB x DBA/2		G_1	1/149 (0.7)	49
PGC (9.5)	F1-B6CB x DBA/2		G_1	3/66 (4.5)	49
PGC (8.5)	F1-B6CB x DBA/2 [or x CD-1]		G_1	2/117 (1.7)	49
Tail-tip fibroblast	*CD-1*		*M*	*0/164 (0)*	22
	CD-1		M	4/130 (3.0)[c]	22
	F1-C57BL/6 x CD-1		M	1/142 (0.7)[c]	22
	F1-C57BL/6 x J1		G_1/S	8/289 (0.5)	188
	F1-B6C3		G_1/S	3/273 (1.0)	189

Table continued on next page

Table 2. Continued

Cell Type	Genotype[a]	Cell Line	Cell Cycle	Cloning Efficiency (%)[b]	Ref.
Cumulus	*C3H*		*G0*	*0/200 (0)*	76
	C57BL/6		*G0*	*0/413 (0)*	76
	DBA/2		*G0*	*1/308 (0.3)*	76
	129/Sv		*G0*	*6/418 (1.4)*	66,76
	F1-129/Sv x 129/Sv		G_0	4/312 (1.3)	66
	F1-129/Sv x C57BL/6		G_0	5/281 (1.8)	66
	F1-129/Sv x DBA/2		G_0	4/294 (1.4)	66
	F1-129/Sv x JF1		G_0	9/481 (1.9)	66
	F1-B6C3		G_0	23/821 (2.8)	76
	F1-B6C3		G_0	0/133 (0)	15
	F1-B6D2 x JF1		G_0	0/297 (0)	66
	F1-B6D2		G_0	10/509 (2.0)	66
	F1-B6D2		G_0	57/1752 (3.3)	76
Immature Sertoli	*129/Sv*		*G1/S*	*10/199 (5.0)*	66
	F1-129/Sv x 129/Sv		G_1/S	6/96 (6.3)	66
	F1-129/Sv x B6D2		G_1/S	28/288 (9.7)	48
	F1-129/Sv x C57BL/6		G_1/S	24/238 (10.1)	66
	F1-129/Sv x DBA/2		G_1/S	16/178 (9.0)	66
	F1-129/Sv x JF1		G_1/S	33/310 (10.7)	66
	F1-B6D2 x JF1		G_1/S	0/127 (0)	66
	F1-B6D2		G_1/S	9/233 (3.9)	66
	F1-B6D2 and -B6CB		G_1/S	16/446 (3.6)	190
Mature Sertoli	F1-B6D2		G_0	0/59(0)	1
T cell	F1-129/Sv x C57BL/6		G_0	0/44 (0)	75
NKT cell	F1-129/Sv x C57BL/6		G_0	4/272 (1.5)	75
Neural precursor	F1-B6D2		G_1/S	5/42 (11.9)	81
Immature neuron	F1-B6D2		G_0	1/42 (2.4)	81
Mature neuron	F1-B6D2		G_0	2/543 (0.4)[d]	82

[a]Genetic backgrounds are listed alphabetically in order of increasing heterozygosity; [b]Surviving offspring as a proportion of the number of embryos transferred; [c] Serial NT; [d] Most clones were scored for normal development at mid-gestation; Inbred donors are italicized. ICM: Inner cell mass, TE: Trophectoderm, ES: embryonic stem cell, PGC: primordial germ cell, isolated from D8.5, 9.5 and 10.5 genital ridges, respectively, NKT: Natural Killer T cell; B6C3=C57BL/6 x C3H, B6CB= C57BL/6 x CBA, B6D2= C57BL/6 x DBA/2

include operator skills, exact composition of media, sources of oocytes and donor cells, culture conditions, embryo transfer (ET), recipient, and calving management, to mention just a few. Variations in any parameter may have subtle consequences on cloning efficiency; however, these are hard to quantify given the small number of live animals obtained at the end of a typical cloning experiment. Owing to the complexity of the NT procedure, it is practically impossible to standardize all experimental details across different research groups.

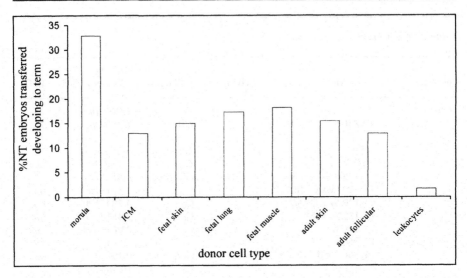

Figure 3. Cloning efficiency and donor cell type in cattle. Small data sets (<10 transfers) from Table 3 were excluded. Data sets from different research groups using the same cell type and cell cycle stage were pooled.

Reprogramming the Donor Cell

After NT of a fully differentiated donor cell, such as a mature T lymphocyte, into an enucleated MII oocyte, the resulting reconstructed embryo can divide and develop into a blastocyst and even a viable animal. The opposite logical possibility, namely that the NT reconstruct develops into a population of fully differentiated lymphocytes, has never been observed. This ability of the oocyte cytoplasm to override whatever transcriptional programme is present in the donor cell has been termed nuclear reprogramming.[9] The reason for this molecular dominance of the oocyte over any other cell type tested is unknown. It may simply be due to the oocyte being three orders of magnitude larger in volume and therefore containing a thousand fold more oocyte-specific factors than the donor cell, in which case the reprogramming dominance should disappear once cell size differences are experimentally adjusted. Alternatively, it may be based on the quality, not the quantity, of those oocyte-specific factors whose natural function is to reprogram the incoming sperm genome after fertilization. In that case, understanding the molecular events of sperm reprogramming may hold a key to understanding, and ultimately improving, the reprogramming of an incoming somatic nucleus as it is more than likely that the reprogramming events and factors in both cases will be the same.

Epigenetic Differences between Donor Cells

Based on the developmental status of the nuclear donor cell, there are three broad categories of cloning: embryonic cell nuclear transfer (ECNT), encompassing blastomeres and embryonic stem cells as donors, germ cell nuclear transfer (GCNT) and somatic cell nuclear transfer (SCNT), using germ or somatic cells of varying differentiation status, respectively, from fetal, newborn or adult sources. Differentiation is a highly regulated process whereby cells become specialized to perform specific functions and loose the ability to perform others. This process is poorly understood in structural terms but appears to be linked to a successive restriction in chromatin accessibility and consequently reduced number of expressed genes.[10,11] Dedifferentiation is the reverse process which leads to the loss of lineage-specific markers, reexpression of nonlineage-specific genes and the associated regaining of proliferative activity. In mammals, dedifferentiation naturally occurs during carcinogenesis and regeneration, which are both relatively rare processes. Experimentally, it requires radical manipulations, such as NT, to induce dedifferentiation.

A hypothesis has been emerging that the donor cell differentiation status is inversely correlated with its cloning efficiency.[12,13] If this were true, the most differentiated cells would be the most difficult to clone, and cloning efficiency after NT could be used as a functional assay to measure varying degrees of differentiation. On the other hand, there may be no relationship between differentiation and reprogramming after NT at all, in which case each donor cell type that requires reprogramming should result in more or less the same cloning efficiency. The following section describes, in order of increasing differentiation, the donor cell types that have been systematically compared in mammalian NT, mainly using mice. It discusses how the results fit the hypothesis and whether some ideal donors are beginning to emerge from the comparison.

Embryonic Cell NT

Blastomeres

The ideal donor genomes would be those that require no reprogramming because their developmental program is identical to that of the recipient oocyte. Male and female pronuclei in the G_1 stage of their first zygotic cell cycle come closest to fulfilling this criterion. The paternal nucleus has already undergone several hours of epigenetic reprogramming including DNA demethylation, the replacement of sperm-specific protamines with histones and progressive post-translational histone modifications, such as hyperacetlyation and methylation (reviewed in ref. 14); early epigenetic reprogramming of the maternal DNA is less well characterized. These remodeling events restore totipotency in the gamete genomes, presumably preparing them for the onset of transcription from the embryonic genome. They are likely to be repeated to some degree after NT of both pronuclei into an MII oocyte. Carrying out such an NT experiment with bovine zygotes is technically difficult because they contain so many lipid droplets that the pronuclei cannot be visualized easily. Fortunately, it has been done recently in the mouse and the results set a benchmark for what amount of improvement in cloning efficiency we can reasonably expect from methodological advances in the future.[15] The answer is, disappointingly, perhaps very little. The authors compared the cloning efficiency of embryos derived from pronuclear NT (PNT) into enucleated MII oocytes with PNT into enucleated zygotes derived from natural mating and with in vitro fertilized (IVF) embryos. Assuming that pronuclei require no reprogramming, any differences must be due to either the use of different recipient cells (MII oocyte vs. zygote) or to the whole micromanipulation procedure itself (PNT vs. IVF). There was no difference between in utero survival of embryos from PNT into zygotes vs. IVF (52% vs. 49%, respectively), showing that the NT procedure per se is not a problem. The slight 1.5 fold difference between embryos derived from PNT into either MII oocytes or zygotes (34% vs. 52%, respectively) was also not significant, indicating that there is little room for technical improvement. In order to detect reprogramming defects right at the onset of embryonic genome activation, it would be very informative to determine the cloning efficiency after PNT from a fertilization-derived zygote into an NT-derived (rather than into another fertilization-derived) zygote, ideally after using different donor cell types to generate NT-derived zygotes.

All initial successes in mammalian cloning were achieved with nuclei from early cleavage-stage embryos of sheep and cattle.[4,5] Following NT in mouse, there appears to be a progressive restriction in cloned embryo development from the 1-cell stage onwards (34% vs. 28% vs. 17% vs. 18% to term after NT of 1-, 2-, 4- and 8-cell nuclei, respectively; Table 2).[15-17] These comparative studies imply that blastomeres become more difficult to reprogram as development proceeds, but raise the question why such reprogramming differences would arise between cells that are still equally totipotent on their own.[18-20] Overall, cloning efficiencies with such early embryonic donor cells are one order of magnitude higher than with somatic cells, as exemplified by using blastomere vs. cumulus cell nuclei (34% vs. 3%), using the same NT technique[15] or mitotic 4-cell vs. fetal fibroblast nuclei (43% vs. 3%), using the same serial NT technique but different genetic backgrounds.[21,22]

Morula formation marks the beginning of cellular differentiation during preimplantation development, eventually leading to the foundation of the trophectoderm and inner cell mass (ICM) lineages from the outer and inner cells of the morula, respectively. Donor cells derived from such later preimplantation stage embryos yield a progressively lower cloning success. However, compared to adult somatic cells, bovine morula blastomeres still are about five-fold more efficient (33% vs. 5%) using slightly different genetic backgrounds and recipient cytoplasts.[23] Fresh ICM cells were first successfully used in cattle,[24,25] soon followed by cloning from short-term bovine ICM cultures.[26] In mouse, successful cloning from ICM and trophectoderm cells required serial NT and revealed no differences in cloning efficiencies between these two cell types (11% vs. 8%).[27] Although abnormal phenotypes are still observed with blastomere cloning (e.g., oversized fetuses and higher perinatal loss), the incidence and severity are greatly reduced, especially after the first trimester.[23] The high success rate with blastomere nuclei is likely due to blastomeres retaining an epigenotype that is more compatible with early embryonic development.

Embryonic Stem (ES) Cells

ES cells are immortal pluripotent (able to give rise to all cell types of the embryo proper) cells derived from placing ICM or morula blastomeres into in vitro culture.[28-30] All attempts to derive livestock ES cells proven to generate germline chimaeras have failed so far. However, long-term "ES-like cell" cultures in farm animal species resemble murine ES cells in morphology, antigen profile, extensive in vitro differentiation ability even after long-term culture and somatic cell chimerism.[31-33] Cloning data with such ES-cell like cells are limited but efficiencies appear to be relatively high,[34] prompting us to predict that a renaissance of embryonic cloning in livestock may be imminent.[35] Cloning efficiencies with murine ES cells can be similarly remarkable as with blastomeres (10-30% vs. 1-3% with fibroblasts or cumulus cells)[36,37] but this is usually associated with the presence of the particular 129 genotype (Table 2). In practice, ES and somatic efficiencies are often similar, even in laboratories that are able to clone from both donor types (1-6%).[38-40] Like the ICM-blastomeres they are derived from, ES cells may retain an epigenotype that is more compatible with early embryonic development and does not need extensive reprogramming.

Germ Cell NT

Primordial Germ (PG) Cells

PG cells (PGCs), the germ cell precursors, have epigenetic modifications that distinguish them from all other cells, such as extensive genome-wide demethylation leading to erasure of allele-specific methylation of imprinted loci. The absence of imprints results in either biallelic expression or repression of specific imprinted genes.[41,42] Contrary to previous expectations,[43] cloning from imprint-free PGCs did not generate viable offspring but resulted in developmental arrest at mid-gestation stages and abnormal placental development.[44-46] This phenotype is consistent with a complete lack of gene dosage regulation for imprinted genes. If PGCs are isolated from a stage before complete imprinting erasure (8.5-10.5 days postcoitum in mouse), they can be completely reprogrammed after NT and give rise to offspring at similar efficiency as somatic cells in cattle and mouse.[47-49] The latter studies in mouse are good examples for how NT can serve as a functional assay to address specific biological questions, in this case, when exactly during PGC development imprints are erased, resulting in loss of their ability to support development to term.

Embryonic Germ (EG) Cells

When PGCs are cultured in vitro they transform into immortal pluripotent EG cells (EGCs).[50,51] They are sometimes also referred to as ES cells, even though they differ in their expression of some imprinted genes.[52] Bona fide EGCs are only available for mouse and humans and no EGCs proven to generate germline chimaeras have been reported from livestock animals. Previous studies have shown that they share a similar epigenotype with the PGCs they are derived from and undergo comparable epigenetic changes. Consequently, imprint-free EGCs are likely to cause the same problems previously encountered with cloning from PGCs.

Germ Cells

Naturally, spermatozoa should be the gold standard for efficient donor cell reprogramming. Their cellular organization is very different from somatic cells: spermatozoa do not synthesize DNA-, mRNA- or protein, contain no ribosomes, endoplasmic reticulum or Golgi apparatus and only have a single centriole in place of a centrosome. They are postmitotic and contain a highly methylated, protamine-based genome which is six-fold more condensed than in somatic cells.[53] As a benchmark control for SCNT, however, spermatozoa are problematic because they only contain a haploid paternal genome and are able to physiologically activate the egg, possibly through delivery of soluble factors such as PLCζ.[54] To avoid ploidy and imprinting problems, NT would have to be performed into mock-enucleated (i.e., after aspirating a small amount of oocyte cytoplasm) MII oocytes without subsequent artificial activation, similar to intracytoplasmic sperm injection (ICSI). Blastocysts derived from ICSI survive much better after embryo transfer than SCNT embryos both in cattle[55] and mouse,[56] however, the comparison is confounded because two steps of the NT procedure are missing (enucleation and artificial activation). Injection of round immature spermatids into MII oocytes, followed by artificial activation, works significantly less efficient than ICSI but more than SCNT (29% vs. 60% vs. 1-3%, respectively),[56] indicating that reprogrammability of both mature and immature haploid sperm nuclei is higher than that of somatic nuclei. Spermatogenic cells from earlier developmental stages are increasingly less efficient, with diploid secondary spermatocytes averaging around 15% and tetraploid primary spermatocytes at 3% developing into live mice (reviewed in ref. 57) which is not greatly different from some types of somatic cells (see below). As in the case of PGCs, cloning efficiency after NT with progressively differentiated spermatids has been used as a functional assay to determine their respective imprinting status.[58]

Somatic Cell NT

The first four decades of NT experiments since Briggs and King introduced the method[59] were dominated by the idea that cloning success depends on the donor cell being somehow embryonic in nature. This notion changed radically with the first successful cloning of sheep from an established embryonic cell line, derived by culturing the differentiated embryonic disc cells over several passages.[60] Since the cloning of Dolly from a differentiated mammary gland cell,[61] various differentiated somatic donor cell types have been used in mouse and cattle NT.

Adult Stem (AS) Cells

The AS cells are commonly defined as undifferentiated multipotent somatic cells capable of proliferation, self-renewal, and the production of differentiating daughter cells. However, these criteria are not obligate since unipotent stem cells exist (e.g., male and female germline stem cells) and life-long self-renewal is also shown by differentiated cells (e.g., B and T lymphocytes). The AS cells have been described in the adult bone marrow, nervous system, skin, muscle and intestine where they are thought to be responsible for regenerating damage and maintaining tissue homeostasis. Some AS cells may even be capable of differentiating across tissue lineage boundaries, however, studies proposing such apparent "lineage plasticity" remain controversial (reviewed in ref. 62). So far, all attempts to define a unique AS cell-specific transcriptome have failed,[63,64] and AS cells are therefore much more difficult to isolate, purify, and maintain in culture than their embryonic equivalents. Their greater developmental plasticity prompts the question, still unanswered, of how useful they will be for cloning.[12] In the near future, comparative NT experiments with AS donor cells could help to better define functional correlates for differentiation and the epigenomic status of these elusive cells.

Progenitor Cells

The difference between progenitors and AS cells is largely semantic, with progenitors being operationally defined as transiently proliferating oligopotent or unipotent cells that have passed the critical transition point of commitment to one or multiple lineages. Freshly isolated neural progenitors from the ventricular side of the cerebral cortex of mouse fetuses have resulted in considerably higher cloning efficiencies than with most other somatic cell type in this species, (12% vs. 2% with putative embryonic neurons and commonly used fibroblasts or cumulus cells), suggesting a correlation between the differentiation and the cloning efficiency, as postulated above. However, numbers were too small and the in vivo results nonsignificant. Immature Sertoli cell progenitors are another type of well-defined donors. Sertoli cells are a major somatic cell type in the testis and central to male gonad formation and spermatogenesis. The conversion from immature, proliferating progenitors to terminally differentiated, nonproliferating Sertoli cells is evident through both morphological and protein markers.[65] In mice, immature Sertoli cells are the most efficient somatic donors by far, achieving up to 15% cloning efficiency with some genotypes[66] and matching the high cloning efficiency of some ES cells. Mature Sertoli cell have not given rise to any offspring.[1]

Fibroblasts are the most commonly used donor cells in cattle cloning and also fit the progenitor-definition. They are generally regarded as unspecialized mesenchymal cells, capable of extensive proliferation, wound healing and differentiation into a variety of other connective tissue lineages, such as bone, cartilage, fat, and muscle. Fetal fibroblasts even contain subpopulations of Oct4-positive cells which can be expanded under appropriate culture conditions and contribute extensively to the mesodermal lineage in chimera experiments.[67] Molecular markers and a unique transcriptional profile for fibroblasts have not been found.[68] Instead, "fibroblasts" which have been traditionally defined by their spindle-shape morphology upon adhering to culture vessels, comprise a host of distinct differentiated cell types, each displaying their own characteristic transcriptome, depending on their site of origin (e.g., lung, skin etc.).[68] For cattle NT, fibroblasts are usually isolated from whole fetuses, or from fetal or adult organs, such as muscle, lung or skin (reviewed in ref. 69). There is no conclusive evidence that the variability in cloning efficiency between cell lines derived from fetal vs. adult sources is any greater than the variability between different cell lines derived from the same source.[69] Likewise, there is no conclusive evidence that the variability between cell lines from the same tissue but different genotypes is any greater than the variability between cell lines from different tissues of the same genotype. Thus, while some donor tissues may be beneficial for in vitro development (e.g., muscle over lung),[70] there is no clear recommendation which fibroblast type to use for cattle cloning (Table 3 and Fig. 3). There is also no indication as to how fibroblasts will compare to cloning efficiencies from cattle ES cell-like donors. In mouse, fibroblasts are rarely used for NT, mainly owing to their large size that interferes with direct microinjection, and cloning efficiencies are lower than with ES cells (Table 2).

Terminally Differentiated Cells

The notion of "terminally" differentiated cells has become somewhat obsolete with the advent of SCNT. While it is certainly true that cells that have reached the differentiation endpoint of their lineage can remain in this stage for the whole life-span of an individual, it has also become clear that even this most differentiated state is not terminal and can be reversed by NT. All initial SCNT experiments, including the cloning of Dolly, have suffered from a lack of accurate donor cell characterization which made it difficult to rule out the possibility that surviving clones were derived from stray somatic stem cells present in the adult donor population.[71] The equal cloning efficiency between nontransgenic and numerous, independently derived, transgenic clonal strains from the same fibroblast donor cell line, already argues against this stem-cell scenario[72] since this could only be explained by postulating that transgene

Table 3. Cloning efficiency and donor cell type in cattle

Cell Type	Cell Cycle	Cloning Efficiency (%)[a]	Reference
4-cell	G_1/S	0/5 (0)	5
8-cell	G_1/S	0/6 (0)	5
9-15-cell	G_1/S	2/7 (28.6)	5
Morula	G_1/S	22/67 (32.8)	23
ICM (fresh)	G_1/S	4/26.9 (15.3)	24
	G_1/S	2/15 (13.3)	25
ICM (cultured)	G_1/S	4/34 (11.8)	26
ES-like	G_1/S	3/7 (42.8)	34
fetal skin	G_0/G_1	6/40 (15.0)	23
	G_0	1/7 (14.3)	191
fetal lung	G_0/G_1	55/318 (17.3)	72
fetal muscle	G_0	4/22 (18.2)	Oback et al, unpublished
adult skin	G_0	4/27 (14.8)	191
	G_0/G_1	26/157 (16.6)	Wells et al, unpublished
	G_0	17/119 (14.3)	6 & unpublished
adult follicular	G_0	36/280 (12.9)	135 & unpublished
Leukocytes	G_0/G_1	2/121 (1.65)	124,192

[a] Births as a proportion of the number of embryos transferred. ICM: Inner cell mass; ES: embryonic stem cell

integration exclusively occurs into stem cells dispersed within the fibroblast population. However, as fibroblasts and putative contaminating stem cells are only poorly defined, it remains a formal possibility. To solve the issue, cloned animals derived from donor nuclei that are genetically marked through terminal differentiation were needed. This was achieved by cloning from mature lymphocytes and neurons.

The B and T lymphocytes are genetically traceable because they undergo genetic rearrangements at the immunoglobulin and T cell receptor loci. Both cell types were used as donors with the B cell being ten times more efficient in generating cloned mice; however, this might have been due to their different genetic background.[73] Cloning success required the use of an indirect two-step procedure, whereby SCNT first generates cloned blastocysts, from which ES cells (ntES cells) are derived for injection into tetraploid host blastocysts in a subsequent round of tetraploid complementation. In this approach, the embryo presumably derives from a very small number of founder ntES cells,[74] whereas the placenta comes mostly from the tetraploid host cells; leaving open the theoretical possibility that the lymphocyte genome was not reprogrammed into all extra-embryonic tissues. Animals from such ntES cell/tetraploid embryo aggregation have been called clones, although, strictly speaking, they are not. The question of whether highly differentiated adult cells are truly totipotent was finally resolved when the first viable mice were cloned from natural killer T (NTK) lymphocytes using conventional single-step NT,[75] confirming that, in principle, cloned mammals could all have been derived from differentiated donor cells. This study compared NTK to helper T cells and confirmed earlier results that cloning from T lymphocytes can be extremely inefficient.[73,76] The two lymphocyte populations were the best characterized donors used to-date, fulfilling all the stringent identification criteria outlined for comparative NT experiments.[77] Cells were isolated from the same hybrid strain of male mice, purified using fluorescence-activated cell sorting based on their surface antigen expression, and determined to be more than 98% pure by flow cytometry. It is encouraging that this positive selection scheme did not seem to compromise their developmental potential for cloning. The NT was performed with fresh cells in the G_0/G_1-cell cycle stage. As a result, this is the first study that conclusively compared cloning efficiency of two epigeneti-cally distinct donor cell subpopulations within the same somatic lineage. Interestingly, NTK cells supported both preimplantation (71% vs. 12% to morula/blastocyst, respectively) and early postimplantation development (60% vs. 7% to implantation) much better than helper T cells. How to interpret these results? NTK cells have been proposed to be a novel lymphocyte subtype characterized by the expression of a single invariant $V\alpha14$ antigen receptor and the NK1.1 marker.[78] They appear to be functionally distinct from conventional $\alpha\beta$ T cells but belong to the same hematopoietic lineage and undergo similar DNA rearrangements to express the T-cell receptor.[79] Could this mean that the NTK cells are "less" differentiated than other T cells? Perhaps, and it would be desirable to test this idea with other assays for cell plasticity or gene expression profiling. At any rate, the NT results independently support the notion that $V\alpha14$ NKT cells indeed constitute a separate sub-lineage on their own.

Extrapolating from this finding may also explain why it is possible to clone from some neuronal lineages but not from others. Initial attempts to clone viable offspring from adult mouse cerebral cortex have repeatedly failed,[1,80] despite the success with embryonic neural progenitors and even young embryonic neurons.[81] Until now, no more than a few normal looking mid-gestation fetuses have been cloned from adult cortical neurons using conventional NT procedures, though not for want of trying by highly experienced groups.[82] While these results prove that post-mitotic neurons can reenter the cell cycle and direct development to some degree, they also clearly indicate that adult neurons are much less efficiently reprogrammed than their progenitors or immature embryonic counterparts. The first successful cloning from adult nerve cells, namely genetically marked olfactory sensory neurons, required the use of tetraploid complementation which confounds the assessment of totipotency as explained above.[83] Presently, these comparisons have not yet revealed any somatic cell lineage with consistently high cloning efficiencies.

In cattle, the situation is similar. Live offspring were obtained from follicular and oviduct epithelial, uterine, mammary gland, skin, liver, lung and muscle fibroblasts and blood leuko-cytes (reviewed in ref. 12), but rigorous comparative studies as in mouse have not been re-ported. Therefore it is currently impossible to assess whether the observed differences in clon-ing efficiency between donor cell types may not be due to other factors. From a practical point of view, the best source of cells for cattle NT depends on the application. Cloning of progeny-tested bulls, for example, currently restricts the source of donor cells to easily acces-sible adult tissue that can be harvested noninvasively. For transgenic applications, where the cells undergo time-consuming selection steps and extensive propagation in vitro, it may be beneficial to use cells from fetuses or calves since they have a longer proliferative lifespan before reaching replicative senescence.[84]

Genetic Differences between Donor Cells

It is evident that cloned embryos, fetuses and offspring suffer from epigenetic abnormali-ties, such as aberrant DNA-methylation,[85-88] histone-methylation[89] and gene expression.[90-94] These abnormal epigenotypes have yet to be causally connected with a particular cloned phe-notype. However, at least some symptoms of the cloned phenotype must indeed be epigenetic since they are not transmitted from parent to offspring.[95]

Most studies to date have either analysed cloned embryos before transfer or after long-term survival in utero or post-natally. This approach allows either no correlation with cloning out-come or selectively analyses only the few survivors. It does not directly answer the question why so many clones fail at various stages of development. Presently, it is still not known to what degree genetic factors, which are presumably nonreprogrammable, contribute to the death of failing clones. Donor cell choice will therefore also influence potential genetic problems in clones.

Genetic Integrity

An unknown proportion of functional adult neurons are aneuploid[96] which may account for their apparent loss of totipotency after NT.[82] All other cell types tested so far contained enough genetic information to direct development of a new organism, even those that eliminate a portion of their genome during differentiation, such as B and T lymphocytes. Tumor cell lines, on the other hand, usually show karyotypic alterations and embryos reconstructed from those often failed to develop to blastocysts or beyond.[97] The ES cells display a 400-fold slower mutation rate than somatic cells, predominantly through chromosome loss and reduplication,[98] and 20-80% of ES cells in any line are estimated to be aneuploid.[99] It is unknown how much this contributes to developmental defects and death of clones. Embryonic carcinoma cells are very similar to ES cells, however, they harbor oncogenic mutations that interfere with directing postimplantation development.[100] Granulosa cells with high levels of chromosomal abnormalities result in NT embryos with significantly more karyotypic errors than controls.[101] Evidently, such donor-derived chromosomal anomalies can affect cloning success, emphasizing the need for rigorous prescreening of donor cells before NT. Higher aneuploidy has also been observed in cloned blastocysts derived from karyotypically normal granulosa cells, thus they can be introduced by the NT procedure itself.[102] Spontaneous mutations arising during ageing and/or time in vitro could affect cloning efficiencies, however, cells after long-term culture[103] or near the end of their replicative life span have no significantly reduced cloning efficiencies.[104]

The mutation rate in mitochondria is much greater than for nuclear DNA,[105] and this may cause problems with mitochondrial (mt) heteroplasmy, i.e., presence of more than just the maternally inherited types of mitochondria within a cell (discussed in detail in Chapter 8). During mouse NT, tail fibroblasts, and cumulus and Sertoli cells all contribute about equal numbers of mtDNA copies to the oocyte (around 0.5% or 5000 copies).[106] Information on mtDNA copy numbers in other cell types is sketchy but since most animal cells in culture appear to contain approximately 1000-5000 molecules per cell,[107,108] this aspect of donor cell selection is unlikely to have a big impact on cloning success. Moreover, conclusive evidence that heteroplasmy impairs development is missing. The majority of investigated SCNT survivors are homoplasmic or mildly heteroplasmic, however, strongly heteroplasmic individuals may be selected against.

Another aspect related to genetic damage is telomere erosion, the progressive shortening of telomeres as a consequence of cell division in the absence of telomerase activity. There are differences in donor cell telomere length, depending on cell type and time of in vitro culture. Bovine ES-like cells show higher telomerase activity than fibroblasts and, for both cell types, early-passages show higher telomerase activity compared to late passage cells.[109] Consequently, average telomere length ($\approx$20 kb) decreases with age and time in culture for both fibroblasts and very late passages of ES-like cells. Telomere length differences between muscle, oviduct, mammary and skin-derived cells were also inferred in another study,[110] however, clear correlations between donor telomere length and cloning efficiency have not been reported. Since senescent cultured bovine fibroblasts with shortened telomeres can be successfully used as donors,[104] telomerase reactivation and telomere length adjustment can be accomplished after NT. Ideally, telomerase activity and telomere length should be compared between the original donor cells and cells of the same type, rederived under the exact same culture conditions from cloned animals of the same age as the original donors. It remains to be convincingly demonstrated that cellular senescence occurring in vivo has anything to do with organismal ageing, that this could result in cloned offspring with premature onset of ageing and a reduced lifespan and that all this would be affected by donor cell choice.

Genetic Background

Cloned mice offer the advantage of a defined genetic background with many inbred strains and F1-crosses being readily available. The highest cloning efficiencies were achieved with oocytes and donor cells of hybrid mice. Cells from inbred mice seldom support postimplantation development, except when either somatic[76] or ES cells[111] were derived from the 129 genotype. This beneficial effect of the 129 genotype can also be observed in heterozygous animals, where, depending on the cell type, it can significantly improve cloning outcome (Table 2).[66] Post-implantation development and post-natal viability improve dramatically when F1 cells are used.[36,37,76,112] The F1-animals are maximally heterozygous for all genetic loci, especially when they come from two inbred strains, and this hybrid vigor is also a well-known phenomenon in farm animal breeding. In cattle, similar comparative studies as in mouse could be undertaken by using highly inbred animals, such as the Chillingham cattle, as a model system.[113] In practice, however, the genetic background will often be dictated by the application, e.g., for cloning progeny-tested bulls, and a particularly favorable genotype has not yet been identified.

Another genotype peculiarity in mammals is their genetic sexual dimorphism, where males have only one X chromosome while females have two. To avoid unequal gene dosage among males and females, one X chromosome is transcriptionally silenced at the blastocyst stage. The X-chromosome inactivation (XCI) occurs by epigenetic modifications and is random in the embryonic lineage but nonrandom in the extra-embryonic lineage where the paternal chromosome is preferentially shut down. This raises the question whether female donor cells are more difficult to clone from due to a faulty reactivation and subsequent nonrandom XCI of the paternal X-chromosome that is active in 50% of the donor cells. The answer is yes, in principle. While reactivation after NT and XCI can be correctly recapitulated in some surviving female mouse clones,[114] others are clearly deficient, resulting in cloned embryos that contain a mixture of cells with varying degrees of XCI[115] and placentas with preferential XCI of the X chromosome inactive in the somatic donor.[114] Likewise, deceased cloned female cattle show aberrant XCI in internal organs and random XCI in the placenta.[116] In vivo, male 129 ntES cell-derived clones from tail-tip fibroblasts develop significantly better into fetuses than female ntES cells from cumulus cells, even though the cloning efficiencies with the original donor cells (fibroblasts vs. cumulus) are very similar.[117] This is in agreement with data from using male and female tail-tip murine fibroblasts.[76] We found no significant differences in cattle SCNT using age- and tissue-matched full-sib brother-sister pairs, although male clones tend to develop better (D. Wells, unpublished observations).

Recipient Oocytes

While there are over 200 donor cell types to choose from, choices for the recipient cell are much more limited. Comparing oocytes of different developmental stages and maturation treatments has defined a subpopulation that currently works best for cattle SCNT. This subpopulation consists of nonactivated MII oocytes from developmentally competent follicles of slaughtered adult animals. The maturation from MI to MII oocytes can occur in vivo or in vitro and both types have been successfully used for cloning. In mice, naturally ovulated oocytes improve in vivo development after ET compared to superovulated ones,[15] however, using them in mono-ovulatory species, such as cattle, is impractical. As with donor cells, hybrid vigor of recipient oocytes tends to be beneficial for development of cloned embryos.[76,118] Further comparative studies are needed to fully evaluate the influence of oocyte source and maturation method on cattle cloning efficiency.

Prior to NT, the recipient oocyte DNA has to be removed or destroyed without compromising viability and reprogramming potential of the cytoplast. This can be done either by gentle aspiration with a glass needle or by oocyte bisection, both methods resulting in similar cloning efficiency.[119] Both procedures could in principle be combined with chemically-assisted enucleation, a method whereby expulsion of the chromosomes in a pseudo-polar body is induced by various combinations of microtubule-depolymerising and oocyte-activating drugs,[120] potentially increasing ease of operation and throughput.[121,122] An intriguing alternative, which does not remove cytoplasmic and plasma membrane components from the egg, is to physically destroy the chromatin through X-ray irradiation.[123] Obviously, in vivo results from this method will be interesting.

Nuclear Transfer

Bona fide NT (by microinjecting isolated nuclei) and various whole-cell NT approaches (by fusion) have been successfully used and the differences are probably not critical. For microinjection, the donor plasma membrane is mechanically ruptured and the cell ghost injected within a few minutes of isolation. The amount of coinjected donor cytoplasmic components is small compared to fusion approaches, measurably decreasing mitochondrial heteroplasmy.[66] On the other hand, direct injection is technically more demanding and not easily applicable to larger donor cells, such as fibroblasts. Fusion methods work best with larger cells and are by far the most widespread method of NT. In contrast to direct injection, all contents of the donor cell, including plasma membrane, organelles and cytosolic factors are introduced into the cytoplast during fusion. Given that oocytes are about 5-10 fold larger in diameter than the typical somatic donor cell, the donor contents get diluted up to 100-fold in the oocyte plasma membrane and up to 1000-fold in the oocytoplasm, probably minimizing their impact on subsequent development. Direct comparisons between piezo-actuated nuclear injection and electrical fusion in cattle found no significant differences in calving rates between the two methods.[124]

Artificial Activation

Cloning subverts the sperm-mediated fertilization step that would normally lead to physiological activation of the oocyte (discussed in detail in Chapter 9). Since mammalian donor cells are unable to activate the recipient cytoplast, various artificial activation protocols have been employed to mimic the sperm-induced cellular events typically occurring during oocyte activation. The choice of oocyte-activating agents has been largely empirical and their effect is functionally nonequivalent to fertilization.[125,126] Comparative studies in cattle and mouse have so far not found any significant differences in cloning efficiency between different oocyte-activating agents.[124,127]

In Vitro Culture of Cloned Embryos

In cattle, the transfer of early-cleavage stage embryos into oviducts is technically difficult and recipient animals are expensive. Since high embryo wastage during preimplantation development would further increase recipient cost, cloned mammalian embryos are cultured in defined media for various periods of time, usually until reaching the blastocyst stage. Studies in mice have shown that tailoring embryo culture medium to the requirements of the donor cell can significantly improve development.[128] Refined in vitro culture systems, where components change in accordance to the needs of the embryo,[129] allow selecting for bovine NT reconstructs that are competent to complete preimplantation development at rates comparable to IVF embryos. Unfortunately, this does not guarantee a high rate post-implantation survival as still about three times more IVF than NT embryos develop into viable offspring.[69]

Embryo Transfer (ET)

In cattle, 1-4 cloned embryos have been transferred into surrogate mothers in the past, but the trend is now towards single transfers. Since survival rates are comparable, this approach results in a 50-75% cost reduction in producing cloned embryos. The possible effect of embryo transfer (ET) has not been conclusively addressed in cattle. In order to do so, in vivo developed single blastocysts derived from natural matings or artificial insemination (AI) would have to be isolated, retransferred and their post-implantation development compared to in vivo developed blastocysts that have not undergone ET. This is a logistically difficult experiment in cattle, which are mono-ovulatory; however, it would not be confounded by other parameters (e.g., superovulation).

Pregnancy Monitoring and Progeny Production

A failure of the placenta to develop and function correctly is a common feature among cattle clones.[130] Although initial day 50 pregnancy rates in cattle following the transfer of single NT embryos can be as high as 65%, and similar to both in vitro fertilised embryos and following AI,[131] there is continual loss thereafter with clones. From our experience, only 13% of cloned embryos transferred result in calves delivered at full term.[132] The magnitude of this pregnancy failure is in stark contrast to the 0-5% loss post-day 50 with AI or natural mating.[133] Attempts are being made to identify molecular markers that detect abnormal placental or fetal development much earlier, allowing elective abortion to lessen the ethical burden associated with the technology. Potential markers may be determined by measuring specific components in the placental fluids[134] or in maternal serum and complemented by detailed ultrasonography. Pregnancy specific protein b produced by the binucleate cells of the trophoblast was transiently higher at day 35 in those concepti that failed to develop to day 90.[130] Similarly, pregnancy serum protein 60 was elevated over the first four months of gestation in those pregnancies that became pathological.[23]

Recipients pregnant with clones generally show poor preparation for parturition and prolonged gestation, with an increased risk of dystocia from heavier birth weight offspring; often prompting elective caesarean section.[135] However, corticosteroid therapy to induce parturition one week before expected full term has successfully aided fetal maturation, (assisted) vaginal delivery and improved the maternal response towards rearing offspring.[136]

The viability of cloned calves at delivery and up to weaning is reduced compared to normal calves and this is despite the greater veterinary care. Data from our group show that around 80% of cloned calves delivered at term are alive after 24 hours, with an additional 15% of calves dying before weaning.[132] Common mortality factors are attributed to dystocia, abnormalities of the cardiovascular, musculoskeletal and neurological systems, as well as susceptibility to gastroenteritis, umbilical and respiratory infections and digestive disorders.[132,137,138]

Adult Clone Phenotypes

The proportion of cloned calves born that are long-term survivors ranges between 47-80%.[132,139-141] Although there are reports that some cloned animals are physiologically normal, at least for the tests examined,[132,140] and may display normal behaviour,[140,142,143] growth rates,[144] reproduction,[1,140] livestock production characteristics,[132,141,145] and life spans,[146] other reports indicate health concerns during the juvenile and adult phases of life. This emphasises the need for detailed long-term follow-up studies of cloned animals. However, the incidence of these clone-associated phenotypes may vary according to the particular species, genotype, sex, cell type or specific aspects of the NT, and culture protocols used. For instance, there is evidence of obesity in cloned mice,[143,147] especially in agouti mouse strains,[146] but, at this stage, there is no indication for early onset obesity occurring in cloned livestock.

A commonly noted clone phenotype is evidence of a compromised immune system. Thymic aplasia has been documented in cloned cattle.[148] Lower levels of antibody production in cloned mice[149] and of cytokines in cloned pigs[150] are direct indicators of a reduced immune response. This may increase their susceptibility to infection and disease, and certainly the incidence of enteritis, umbilical and respiratory infections are increased in cloned livestock. Others, however, have reported normal characteristics of peripheral blood lymphocytes and normal responses to periodic infection in cloned cattle.[140]

The majority of mice cloned from immature Sertoli cells of the B6D2 hybrid strain died after approximately 500 days which was around 50% of the lifespan in control mice.[149] The causes of death included severe pneumonia and hepatic failure. It remains to be determined whether this is a general phenomenon with clones, but it appears that this phenomenon is both cell type and genotype specific, with other cloned mice having apparently normal life spans.[146]

The mouse has the advantage of a shorter generation interval and biological lifespan to screen for these effects. While it is encouraging that some studies report normal health of four year old bovine clones.[140] it may be too early to detect the phenotype reported in mice. Studies at AgResearch show that between weaning and four years of age, the annual mortality rate in SCNT cattle is at least 8%.[132] This is in marked contrast to the negligible mortality experienced with the offspring of clones and the typically accepted mortality of 2-3% per annum in conventional free range farming. Although the reasons for death among the clones are varied, and some potentially preventable, the main mortality factor beyond weaning is euthanasia due to musculoskeletal abnormalities.[132] This includes cases of chronic lameness in milking cows and emphasises that any underlying frailties in clones may not be fully revealed until the animals are stressed in some manner.

The cloned offspring syndrome is a continuum, in that lethality or abnormal phenotypes may occur at any phase of development presumably depending upon the degree of dysregulation of key genes arising from fundamental errors in epigenetic reprogramming. Epigenetic aberrations that occur early in embryonic or foetal development may impair health in adulthood and cause some clone-associated phenotypes.[146] Even apparently normal clones may have abnormal regulation of many genes that are too subtle to result in an obvious phenotype.[92]

Trans-Generational Effects

It is important to not only monitor the health of the clones but also their subsequent progeny derived following sexual reproduction. Offspring of male and female clones in a range of species have been produced following both natural mating and assisted sexual reproduction, such as AI, with a noncloned partner. Conception, pregnancy, parturition and survival rates are all within normal limits,[1,132,140] as is the subsequent fertility of these offspring of clones (Oback and Wells, unpublished). More discriminatory is the mating of cloned females with cloned males. With these matings in sheep,[136] cattle (Oback and Wells, unpublished) and mice[147] there is no evidence of the placental abnormalities and large birth weights recorded in the clones' offspring. It has also been claimed that the obese phenotype observed in cumulus cell mouse clones is not heritable following mating with cloned males derived from fibroblasts of the same mouse strain.[147] However, it has not been reported whether the males of this strain also have the obese phenotype. If the obesity is truly nonheritable, then another generation of inbreeding would be required to exclude the possibility of a recessive genetic (or epigenetic) trait. The most convincing evidence for the lack of transmission of any obvious deleterious recessive genetic or epigenetic trait has been provided following the mating of cloned male and cloned female mice derived from XY and XO ES cells, respectively, obtained from the same cell line.[95] The resulting offspring were phenotypically normal, lacking the fetal and placental overgrowth and open eyelids at birth characteristic of their cloned parents. However, these cloned parents were themselves the survivors of a cloning experiment. Although additional breeding studies would be necessary to exclude the presence of any recessive (epi)genetic traits, it is interesting that normal progeny have resulted from ICSI following the grafting of testicular tissue from cloned mice that died at birth.[151]

These observations indicate that clone-associated phenotypes are epigenetic and any errors in the surviving clones appear to be corrected during gametogenesis. This provides some confidence for the major application of cloning technology in agriculture; namely, the generation of cloned sires for breeding. Nonetheless, it is still possible that genetic errors may be present in the clones that would be heritable. Moreover, detailed molecular studies are required to confirm whether the necessary epigenetic modifications in gametes, zygotes and embryos derived from cloned parents are indeed restored to normal. It is critical to investigate this phenomenon more thoroughly, as evidence exists for the germline transmission of epigenetic states at various endogenous loci[152,153] and in more artificial situations, following nuclear-cytoplasmic incompatibility.[154]

Applications of Cattle Cloning

There are several important issues to be addressed before commercial opportunities for cloning in livestock agriculture can be realised. Animal welfare concerns limit acceptability and applicability of the technology in its current inefficient form. There needs to be confidence in the long-term health status of cloned livestock and progeny in subsequent generations. Commercialisation is awaiting regulatory approval on the safety of food products derived from clones and their offspring. Even with regulatory approval, industry, farmers and consumers need to accept the technology. Some of the cattle cloning applications discussed below will not be realised until well into the future.

Multiplying Valuable Genotypes

Cloning could enable the rapid dissemination of superior genotypes from nucleus breeding herds, directly to commercial farmers. Genotypes could be provided that are ideally suited for specific product characteristics, disease resistance, or environmental conditions. Cloning could be extremely useful in multiplying outstanding F_1 crossbred animals, or composite breeds, to maximise the benefits of both heterosis and maintaining favourable combinations of alleles in heterozygotes, that would otherwise segregate in the F_2 generation. The dissemination of genetic gain could be achieved through the controlled release of selected lines of elite cloned embryos or, most appropriately, by the production of cloned animals with superior genetics for breeding. These could be clones of progeny-tested animals, especially sires. This is more relevant in the beef industry, where cloned sires would provide an effective means of disseminating their superior genetics through widespread natural mating, effectively substituting for AI which is expensive and inconvenient in more extensive farming systems. Even in the dairy industry, cloned bulls could generate extra semen to supply unmet international demand for insemination doses from top sires. To even more rapidly disseminate genetic gain, and reduce genetic lag by at least two generations, it will be advantageous to clone from embryonic blastomeres following marker assisted genetic selection to identify superior embryos, rather than using somatic cells from adult animals.[35]

Preservation of Endangered Cattle Breeds

Cloning can be used to help preserve indigenous or traditional breeds of livestock that have production traits and adaptability to local environments that should not be lost from the global gene pool.[155] This is a very important application as most of the genetic variation in a livestock species resides in the interbreed variability.[156] Thus, their demise represents a significant loss of biodiversity and limits future opportunities to capture as yet unappreciated traits. More imperative than cloning per se is the cryopreservation of somatic cells from rare breeds with the future prospect of cloning deceased animals and reintroducing their genetics back into the live breeding population.[157] Cryobanking of somatic cells as potential future NT donors thus provides an insurance policy against further losses of diversity or possible extinction and would be easier than preserving gametes or embryos. Even for conventional agriculture, it might be prudent to cryopreserve somatic cells from genetically elite animals in case of disease or accidental death.

Research Models

Sets of cloned livestock animals could be used to reduce genetic variability and hence also the numbers of animals needed in some experimental studies. This could be conducted on a larger scale than what is currently achieved with sets of monozygotic twins[158] but accepting the greater mitochondrial DNA and environmental differences with NT clones. Lambs cloned from sheep selected either for resistance or susceptibility to nematode worms will have utility in studies aimed at discovering novel genes and regulatory pathways in immunology.[159] However, the use of clones as research models needs to account for the greater than expected phenotypic variability for some traits, due to differences in epigenetic reprogramming.[160] Cattle have been well used as models to address the histocompatibility of cell based therapies in proof-of-principle studies.[161,162]

Cloning for Transgenic Applications

A major application is to clone animals from cells that have been genetically modified in order to produce transgenic livestock. The NT is only one of several of methods available to produce transgenic animals.[163] However, despite the current problems with reprogramming, the approach of cell-mediated transgenesis has a number of advantages, including the ability to:

a. Introduce,[164] functionally delete,[165] or subtly modify genes of interest
b. Introduce a specific transgene into a desired genetic background of the chosen sex (particularly important for agricultural traits)
c. Screen cells for the specific genetic modification before producing the transgenic animal
d. Produce embryos or offspring that are all transgenic and where none should be mosaic (with a mixture of transgenic and nontransgenic cells in the same organism)
e. Produce small herds from each cell line in the first generation, rather than individual founder animals that need to be subsequently bred.

While SCNT can produce a few founder transgenic animals, it is presently still desirable to use assisted sexual reproduction thereafter to further multiply animals without the potential epigenetic aberrations of clones.[166] The most efficient means of introducing a transgene into the wider livestock population is through AI or natural mating. Ideally, the sire should be homozygous for the desired trait so all progeny receive a copy of the transgene. Animal industries may choose to annually introduce the same transgene on a new genetic background using cell lines derived from the most recently selected progeny tested sires, so as to capture the annual incremental genetic gains from conventional animal breeding. The economic benefits of a genetic modification affecting a production trait must be sufficiently large to compensate for the lag in genetic gain during the time taken to introduce the transgene and test its performance in a variety of genotypes, as phenotypic outcome may differ,[167] before wider dissemination.

There are a range of opportunities available for genetically modifying cattle for both biomedicine and agriculture, depending upon the particular genes that are altered. There has been greater research effort, progress, economic incentive, ethical justification, and public acceptance towards the generation of transgenic livestock for various biomedical applications, compared to those for agriculture and food production. However, the opportunities for biomedicine, especially in the areas of xenotransplantation, and models for human genetic diseases, are considerably more limited in cattle compared to smaller livestock species, due to their greater size, expense and generation intervals.

A major opportunity is biopharming. Transgenic females have been cloned from cells containing medically important human genes under the control of promoters that direct expression to the lactating mammary gland.[164,168] Following secretion into the milk, these therapeutic proteins may then be extracted from milk, purified, and used in clinical trials to evaluate their safety and effectiveness in treating human diseases and disorders before gaining regulatory approval. Livestock are favored wherever functional proteins are difficult to make in sufficient quantities, cost-effectively, and safely by other methods. In harnessing the potential of the mammary gland to synthesise heterologous proteins, the choice of species (from rabbits to cows) depends upon the quantities required.[169] Indeed, other tissue systems may sometimes be preferable; for instance, antibodies produced in the eggs of chickens[170] or blood of transgenic cattle.[171]

As understanding of the genes that influence livestock production traits improves, so does the knowledge of how to accurately modify the appropriate genes and regulatory sequences to generate new and desired animal products in the future. Agricultural applications of transgenesis in cattle are primarily aimed at increasing the quantity and quality of valuable meat and milk components and improving environmental sustainability that will have economic benefits for farmers and processors, or additional health benefits for consumers.

Genetic modification of milk composition in dairy cattle has received considerable attention in efforts to improve production, aid human nutrition and alter various processing properties designed to suit the manufacturing of specific food products.[172,173] Manipulations include either the over-expression of endogenous genes or the introduction of a foreign gene producing a novel protein in milk to generate a nutraceutical (medical food). The introduction of additional copies of bovine β- and κ-casein genes into cloned dairy heifers resulted in at least a two-fold increase in κ-casein protein content in milk within one generation.[174] It remains to be determined how this affects the processing properties for cheese manufacture. It is important to consider that particular transgenic milk streams, tailored for specific purposes, might be unsuitable for general commodity milk products. One possible scenario for the future is the generation of herds possessing specific genetic modifications producing agricultural products for niche markets. In the dairy industry, transgenic milk from specific herds would need to be kept separate for manufacturing purposes, let alone for food labelling compliance. Such a prospect will pose challenges for the structure of traditional commodity-based dairy industries processing bulk milk. The integration of transgenesis might necessitate regional herds producing milk of a similar type with specific processing capability available locally.

There is great appeal in the prospect of using transgenesis in livestock to aid sustainable agriculture. The intention to improve animal health, reduce pollution, and more effectively utilise both feed and animal resources might be better received by society. Although improved disease and pest resistance has long been an ambitious goal,[175] the result would be improved animal welfare and reduced reliance on animal remedies. By extrapolation from the mouse,[176,177] either inactivation[178] or introduction of a mutated PrP gene in livestock[179] would be expected to produce animals resistant to prion diseases. Cattle resistant to transmissible spongiform encephalopathies would be especially valuable in providing improved safeguards for biomedical applications. There has been success in the prevention of mastitis caused by *Staphylococcus aureus* infection in dairy cattle, following the introduction of a biologically active form of lysostaphin,[180] and resistance to ectoparasites may be possible with the production of chitinase in the skin to kill larvae.[181] Bacterial genes encoding the components of biochemical pathways that are nonfunctional in livestock may increase the feed utilisation efficiency from dietary roughage,[182] the bioavailability of specific nutrients from animal feed and reducing environmental pollution,[183] or enable de novo synthesis of essential amino acids that are otherwise rate limiting for production.[182] The prospect of generating cloned bulls from progeny-tested sires that have been genetically modified to produce sperm only of a single sex[184] would be highly desirable. In the dairy industry, sires only producing X-chromosome bearing sperm would allow greater selection intensity on dams for replacement daughters and avoid an excess of unwanted bull calves. Conversely, in the beef industry there would be a preference for sires only producing male offspring. Another advantage for market acceptability of this strategy is that the offspring would not themselves be transgenic.[184]

Acknowledgements

This work was supported by the New Zealand Foundation for Research, Science and Technology and AgResearch.

References

1. Wakayama T, Perry AC, Zuccotti M et al. Full-term development of mice from enucleated oocytes injected with cumulus cell nuclei. Nature 1998; 394(6691):369-374.
2. Memili E, First NL. Zygotic and embryonic gene expression in cow: A review of timing and mechanisms of early gene expression as compared with other species. Zygote 2000; 8(1):87-96.
3. McGrath J, Solter D. Nuclear transplantation in the mouse embryo by microsurgery and cell fusion. Science 1983; 220(4603):1300-1302.
4. Willadsen SM. Nuclear transplantation in sheep embryos. Nature 1986; 320(6057):63-65.

5. Prather RS, Barnes FL, Sims MM et al. Nuclear transplantation in the bovine embryo: Assessment of donor nuclei and recipient oocyte. Biol Reprod 1987; 37(4):859-866.
6. Oback B, Wiersema AT, Gaynor P et al. Cloned cattle derived from a novel zona-free embryo reconstruction system. Cloning Stem Cells 2003; 5(1):3-12.
7. Peura TT, Lewis IM, Trounson AO. The effect of recipient oocyte volume on nuclear transfer in cattle. Mol Reprod Dev 1998; 50(2):185-191.
8. Vajta G, Lewis IM, Hyttel P et al. Somatic cell cloning without micromanipulators. Cloning 2001; 3(2):89-95.
9. Western PS, Surani MA. Nuclear reprogramming—alchemy or analysis? Nat Biotechnol 2002; 20(5):445-446.
10. Kramer JA, McCarrey JR, Djakiew D et al. Differentiation: The selective potentiation of chromatin domains. Development 1998; 125(23):4749-4755.
11. Muller C, Leutz A. Chromatin remodeling in development and differentiation. Curr Opin Genet Dev 2001; 11(2):167-174.
12. Oback B, Wells D. Donor cells for nuclear cloning: Many are called, but few are chosen. Cloning Stem Cells 2002; 4(2):147-168.
13. Jaenisch R, Eggan K, Humpherys D et al. Nuclear cloning, stem cells, and genomic reprogramming. Cloning Stem Cells 2002; 4(4):389-396.
14. Morgan HD, Santos F, Green K et al. Epigenetic reprogramming in mammals. Hum Mol Genet 2005; 14(Spec No 1):R47-58.
15. Hiiragi T, Solter D. Reprogramming is essential in nuclear transfer. Mol Reprod Dev 2005; 70(4):417-421.
16. Cheong HT, Takahashi Y, Kanagawa H. Birth of mice after transplantation of early cell-cycle-stage embryonic nuclei into enucleated oocytes. Biol Reprod 1993; 48(5):958-963.
17. Otaegui PJ, O'Neill GT, Campbell KH et al. Transfer of nuclei from 8-cell stage mouse embryos following use of nocodazole to control the cell cycle. Mol Reprod Dev 1994; 39(2):147-152.
18. Johnson WH, Loskutoff NM, Plante Y et al. Production of four identical calves by the separation of blastomeres from an in vitro derived four-cell embryo. Vet Rec 1995; 137(1):15-16.
19. Modlinski JA, Ozil JP, Modlinska MK et al. Development of single mouse blastomeres enlarged to zygote size in conditions of nucleo-cytoplasmic synchrony. Zygote 2002; 10(4):283-290.
20. Willadsen SM, Polge C. Attempts to produce monozygotic quadruplets in cattle by blastomere separation. Vet Rec 1981; 108(10):211-213.
21. Kwon OY, Kono T. Production of identical sextuplet mice by transferring metaphase nuclei from four-cell embryos. Proc Natl Acad Sci USA 1996; 93(23):13010-13013.
22. Ono Y, Shimozawa N, Ito M et al. Cloned mice from fetal fibroblast cells arrested at metaphase by a serial nuclear transfer. Biol Reprod 2001; 64(1):44-50.
23. Heyman Y, Chavatte-Palmer P, LeBourhis D et al. Frequency and occurrence of late-gestation losses from cattle cloned embryos. Biol Reprod 2002; 66(1):6-13.
24. Keefer CL, Stice SL, Matthews DL. Bovine inner cell mass cells as donor nuclei in the production of nuclear transfer embryos and calves. Biol Reprod 1994; 50(4):935-939.
25. Collas P, Barnes FL. Nuclear transplantation by microinjection of inner cell mass and granulosa cell nuclei. Mol Reprod Dev 1994; 38(3):264-267.
26. Sims M, First NL. Production of calves by transfer of nuclei from cultured inner cell mass cells. Proc Natl Acad Sci USA 1994; 91(13):6143-6147.
27. Tsunoda Y, Kato Y. Not only inner cell mass cell nuclei but also trophectoderm nuclei of mouse blastocysts have a developmental totipotency. J Reprod Fertil 1998; 113(2):181-184.
28. Evans MJ, Kaufman MH. Establishment in culture of pluripotential cells from mouse embryos. Nature 1981; 292(5819):154-156.
29. Martin GR. Isolation of a pluripotent cell line from early mouse embryos cultured in medium conditioned by teratocarcinoma stem cells. Proc Natl Acad Sci USA 1981; 78(12):7634-7638.
30. Tesar PJ. Derivation of germ-line-competent embryonic stem cell lines from preblastocyst mouse embryos. Proc Natl Acad Sci USA 2005; 102(23):8239-8244.
31. Cibelli JB, Stice SL, Golueke PJ et al. Transgenic bovine chimeric offspring produced from somatic cell-derived stem-like cells. Nat Biotechnol 1998; 16(7):642-646.
32. Mitalipova M, Beyhan Z, First NL. Pluripotency of bovine embryonic cell line derived from precompacting embryos. Cloning 2001; 3(2):59-67.
33. Wang L, Duan E, Sung LY et al. Generation and characterization of pluripotent stem cells from cloned bovine embryos. Biol Reprod 2005; 73(1):149-155.
34. Saito S, Sawai K, Ugai H et al. Generation of cloned calves and transgenic chimeric embryos from bovine embryonic stem-like cells. Biochem Biophys Res Commun 2003; 309(1):104-113.

35. Wells DN, Oback B, Laible G. Cloning livestock: A return to embryonic cells. Trends Biotechnol 2003; 21(10):428-432.
36. Eggan K, Akutsu H, Loring J et al. Hybrid vigor, fetal overgrowth, and viability of mice derived by nuclear cloning and tetraploid embryo complementation. Proc Natl Acad Sci USA 2001; 98(11):6209-6214.
37. Rideout IIIrd WM, Wakayama T, Wutz A et al. Generation of mice from wild-type and targeted ES cells by nuclear cloning. Nat Genet 2000; 24(2):109-110.
38. Amano T, Tani T, Kato Y et al. Mouse cloned from embryonic stem (ES) cells synchronized in metaphase with nocodazole. J Exp Zool 2001; 289(2):139-145.
39. Ono Y, Shimozawa N, Muguruma K et al. Production of cloned mice from embryonic stem cells arrested at metaphase. Reproduction 2001; 122(5):731-736.
40. Zhou Q, Jouneau A, Brochard V et al. Developmental potential of mouse embryos reconstructed from metaphase embryonic stem cell nuclei. Biol Reprod 2001; 65(2):412-419.
41. Li E, Beard C, Jaenisch R. Role for DNA methylation in genomic imprinting. Nature 1993; 366(6453):362-365.
42. Szabo PE, Mann JR. Biallelic expression of imprinted genes in the mouse germ line: Implications for erasure, establishment, and mechanisms of genomic imprinting. Genes Dev 1995; 9(15):1857-1868.
43. Jaenisch R. DNA methylation and imprinting: Why bother? Trends Genet 1997; 3(8):323-329.
44. Kato Y, Rideout IIIrd WM, Hilton K et al. Developmental potential of mouse primordial germ cells. Development 1999; 126(9):1823-1832.
45. Lee J, Inoue K, Ono R et al. Erasing genomic imprinting memory in mouse clone embryos produced from day 11.5 primordial germ cells. Development 2002; 129(8):1807-1817.
46. Yamazaki Y, Mann MR, Lee SS et al. Reprogramming of primordial germ cells begins before migration into the genital ridge, making these cells inadequate donors for reproductive cloning. Proc Natl Acad Sci USA 2003; 100(21):12207-12212.
47. Zakhartchenko V, Durcova-Hills G, Schernthaner W et al. Potential of fetal germ cells for nuclear transfer in cattle. Mol Reprod Dev 1999; 52(4):421-426.
48. Miki H, Inoue K, Kohda T et al. Birth of mice produced by germ cell nuclear transfer. Genesis 2005; 41(2):81-86.
49. Yamazaki Y, Low EW, Marikawa Y et al. Adult mice cloned from migrating primordial germ cells. Proc Natl Acad Sci USA 2005; 102(32):11361-11366.
50. Matsui Y, Zsebo K, Hogan BL. Derivation of pluripotential embryonic stem cells from murine primordial germ cells in culture. Cell 1992; 70(5):841-847.
51. Stewart CL, Gadi I, Bhatt H. Stem cells from primordial germ cells can reenter the germ line. Dev Biol 1994; 161(2):626-628.
52. Labosky PA, Barlow DP, Hogan BL. Mouse embryonic germ (EG) cell lines: Transmission through the germline and differences in the methylation imprint of insulin-like growth factor 2 receptor (Igf2r) gene compared with embryonic stem (ES) cell lines. Development 1994; 120(11):3197-3204.
53. Ward WS, Coffey DS. DNA packaging and organization in mammalian spermatozoa: Comparison with somatic cells. Biol Reprod 1991; 44(4):569-574.
54. Saunders CM, Larman MG, Parrington J et al. PLC zeta: A sperm-specific trigger of Ca(2+) oscillations in eggs and embryo development. Development 2002; 129(15):3533-3544.
55. Wei H, Fukui Y. Births of calves derived from embryos produced by intracytoplasmic sperm injection without exogenous oocyte activation. Zygote 2002; 10(2):149-153.
56. Kishigami S, Wakayama S, Nguyen VT et al. Similar time restriction for intracytoplasmic sperm injection and round spermatid injection into activated oocytes for efficient offspring production. Biol Reprod 2004; 70(6):1863-1869.
57. Yanagimachi R. Intracytoplasmic injection of spermatozoa and spermatogenic cells: Its biology and applications in humans and animals. Reprod Biomed Online 2005; 10(2):247-288.
58. Kimura Y, Yanagimachi R. Mouse oocytes injected with testicular spermatozoa or round spermatids can develop into normal offspring. Development 1995; 121(8):2397-2405.
59. Briggs R, King TJ. Transplantation of living nuclei from blastula cells into enucleated frogs' eggs. Proc Natl Acad Sci USA 1952; 38:455-463.
60. Campbell KH, McWhir J, Ritchie WA et al. Sheep cloned by nuclear transfer from a cultured cell line. Nature 1996; 380(6569):64-66.
61. Wilmut I, Schnieke AE, McWhir J et al. Viable offspring derived from fetal and adult mammalian cells. Nature 1997; 385(6619):810-813.
62. Raff M. Adult stem cell plasticity: Fact or artifact? Annu Rev Cell Dev Biol 2003; 19:1-22.
63. Evsikov AV, Solter D. Comment on " 'Stemness': Transcriptional profiling of embryonic and adult stem cells" and "a stem cell molecular signature". Science 2003; 302(5644):393, (author reply 393).

64. Fortunel NO, Otu HH, Ng HH et al. Comment on " 'Stemness': Transcriptional profiling of embryonic and adult stem cells" and "a stem cell molecular signature". Science 2003; 302(5644):393, (author reply 393).
65. Sharpe RM, McKinnell C, Kivlin C et al. Proliferation and functional maturation of Sertoli cells, and their relevance to disorders of testis function in adulthood. Reproduction 2003; 125(6):769-784.
66. Inoue K, Ogonuki N, Mochida K et al. Effects of donor cell type and genotype on the efficiency of mouse somatic cell cloning. Biol Reprod 2003; 69(4):1394-1400.
67. Kues WA, Petersen B, Mysegades W et al. Isolation of murine and porcine fetal stem cells from somatic tissue. Biol Reprod 2005; 72(4):1020-1028.
68. Chang HY, Chi JT, Dudoit S et al. Diversity, topographic differentiation, and positional memory in human fibroblasts. Proc Natl Acad Sci USA 2002; 99(20):12877-12882.
69. Oback B, Wells DN. Cloning cattle. Cloning Stem Cells 2003; 5(4):243-256.
70. Powell AM, Talbot NC, Wells KD et al. Cell donor influences success of producing cattle by somatic cell nuclear transfer. Biol Reprod 2004; 71(1):210-216.
71. Liu L. Cloning efficiency and differentiation. Nat Biotechnol 2001; 19(5):406.
72. Wells DN, Laible G, Tucker FC et al. Coordination between donor cell type and cell cycle stage improves nuclear cloning efficiency in cattle. Theriogenology 2003; 59(1):45-59.
73. Hochedlinger K, Jaenisch R. Monoclonal mice generated by nuclear transfer from mature B and T donor cells. Nature 2002; 415(6875):1035-1038.
74. Wang Z, Jaenisch R. At most three ES cells contribute to the somatic lineages of chimeric mice and of mice produced by ES-tetraploid complementation. Dev Biol 2004; 275(1):192-201.
75. Inoue K, Wakao H, Ogonuki N et al. Generation of cloned mice by direct nuclear transfer from natural killer T cells. Curr Biol 2005; 15(12):1114-1118.
76. Wakayama T, Yanagimachi R. Mouse cloning with nucleus donor cells of different age and type. Mol Reprod Dev 2001; 58(4):376-383.
77. Oback B, Wells D. Practical aspects of donor cell selection for nuclear cloning. Cloning Stem Cells 2002; 4(2):169-175.
78. Taniguchi M, Nakayama T. Recognition and function of Valpha14 NKT cells. Semin Immunol 2000; 12(6):543-550.
79. Taniguchi M, Harada M, Kojo S et al. The regulatory role of Valpha14 NKT cells in innate and acquired immune response. Annu Rev Immunol 2003; 21:483-513.
80. Osada T, Kusakabe H, Akutsu H et al. Adult murine neurons: Their chromatin and chromosome changes and failure to support embryonic development as revealed by nuclear transfer. Cytogenet Genome Res 2002; 97(1-2):7-12.
81. Yamazaki Y, Makino H, Hamaguchi-Hamada K et al. Assessment of the developmental totipotency of neural cells in the cerebral cortex of mouse embryo by nuclear transfer. Proc Natl Acad Sci USA 2001; 98(24):14022-14026.
82. Makino H, Yamazaki Y, Hirabayashi T et al. Mouse embryos and chimera cloned from neural cells in the postnatal cerebral cortex. Cloning Stem Cells 2005; 7(1):45-61.
83. Eggan K, Baldwin K, Tackett M et al. Mice cloned from olfactory sensory neurons. Nature 2004; 428(6978):44-49.
84. Kasinathan P, Knott JG, Moreira PN et al. Effect of fibroblast donor cell age and cell cycle on development of bovine nuclear transfer embryos in vitro. Biol Reprod 2001; 64(5):1487-1493.
85. Bourc'his D, Le Bourhis D, Patin D et al. Delayed and incomplete reprogramming of chromosome methylation patterns in bovine cloned embryos. Curr Biol 2001; 11:1542-1546.
86. Dean W, Santos F, Stojkovic M et al. Conservation of methylation reprogramming in mammalian development: Aberrant reprogramming in cloned embryos. Proc Natl Acad Sci USA 2001; 98(24):13734-13738.
87. Kang YK, Koo DB, Park JS et al. Aberrant methylation of donor genome in cloned bovine embryos. Nat Genet 2001; 28(2):173-177.
88. Kang YK, Park JS, Koo DB et al. Limited demethylation leaves mosaic-type methylation states in cloned bovine preimplantation embryos. EMBO J 2002; 21(5):1092-1100.
89. Santos F, Zakhartchenko V, Stojkovic M et al. Epigenetic marking correlates with developmental potential in cloned bovine preimplantation embryos. Curr Biol 2003; 13(13):1116-1121.
90. Boiani M, Eckardt S, Scholer HR et al. Oct4 distribution and level in mouse clones: Consequences for pluripotency. Genes Dev 2002; 16(10):1209-1219.
91. Bortvin A, Eggan K, Skaletsky H et al. Incomplete reactivation of Oct4-related genes in mouse embryos cloned from somatic nuclei. Development 2003; 130(8):1673-1680.
92. Humpherys D, Eggan K, Akutsu H et al. Abnormal gene expression in cloned mice derived from embryonic stem cell and cumulus cell nuclei. Proc Natl Acad Sci USA 2002; 99(20):12889-12894.

93. Pfister-Genskow M, Myers C, Childs LA et al. Identification of differentially expressed genes in individual bovine preimplantation embryos produced by nuclear transfer: Improper reprogramming of genes required for development. Biol Reprod 2005; 72(3):546-555.
94. Wrenzycki C, Wells D, Herrmann D et al. Nuclear transfer protocol affects messenger RNA expression patterns in cloned bovine blastocysts. Biol Reprod 2001; 65(1):309-317.
95. Shimozawa N, Ono Y, Kimoto S et al. Abnormalities in cloned mice are not transmitted to the progeny. Genesis 2002; 34(3):203-207.
96. Kingsbury MA, Friedman B, McConnell MJ et al. Aneuploid neurons are functionally active and integrated into brain circuitry. Proc Natl Acad Sci USA 2005; 102(17):6143-6147.
97. Hochedlinger K, Blelloch R, Brennan C et al. Reprogramming of a melanoma genome by nuclear transplantation. Genes Dev 2004; 18(15):1875-1885.
98. Cervantes RB, Stringer JR, Shao C et al. Embryonic stem cells and somatic cells differ in mutation frequency and type. Proc Natl Acad Sci USA 2002; 99(6):3586-3590.
99. Eggan K, Rode A, Jentsch I et al. Male and female mice derived from the same embryonic stem cell clone by tetraploid embryo complementation. Nat Biotechnol 2002; 20(5):455-459.
100. Blelloch RH, Hochedlinger K, Yamada Y et al. Nuclear cloning of embryonal carcinoma cells. Proc Natl Acad Sci USA 2004; 101(39):13985-13990.
101. Bureau WS, Bordignon V, Leveillee C et al. Assessment of chromosomal abnormalities in bovine nuclear transfer embryos and in their donor cells. Cloning Stem Cells 2003; 5(2):123-132.
102. Booth PJ, Viuff D, Tan S et al. Numerical chromosome errors in day 7 somatic nuclear transfer bovine blastocysts. Biol Reprod 2003; 68(3):922-928.
103. Kubota C, Yamakuchi H, Todoroki J et al. Six cloned calves produced from adult fibroblast cells after long-term culture. Proc Natl Acad Sci USA 2000; 97(3):990-995.
104. Lanza RP, Cibelli JB, Blackwell C et al. Extension of cell life-span and telomere length in animals cloned from senescent somatic cells. Science 2000; 288(5466):665-669.
105. Pesole G, Gissi C, De Chirico A et al. Nucleotide substitution rate of mammalian mitochondrial genomes. J Mol Evol 1999; 48(4):427-434.
106. Inoue K, Ogonuki N, Yamamoto Y et al. Tissue-specific distribution of donor mitochondrial DNA in cloned mice produced by somatic cell nuclear transfer. Genesis 2004; 39(2):79-83.
107. Bogenhagen D, Clayton DA. The number of mitochondrial deoxyribonucleic acid genomes in mouse L and human HeLa cells. Quantitative isolation of mitochondrial deoxyribonucleic acid. J Biol Chem 1974; 249(24):7991-7995.
108. Shmookler Reis RJ, Goldstein S. Mitochondrial DNA in mortal and immortal human cells. Genome number, integrity, and methylation. J Biol Chem 1983; 258(15):9078-9085.
109. Betts D, Bordignon V, Hill J et al. Reprogramming of telomerase activity and rebuilding of telomere length in cloned cattle. Proc Natl Acad Sci USA 2001; 98(3):1077-1082.
110. Miyashita N, Shiga K, Yonai M et al. Remarkable differences in telomere lengths among cloned cattle derived from different cell types. Biol Reprod 2002; 66(6):1649-1655.
111. Gao S, McGarry M, Ferrier T et al. Effect of cell confluence on production of cloned mice using an inbred embryonic stem cell line. Biol Reprod 2003; 68(2):595-603.
112. Wakayama T, Rodriguez I, Perry AC et al. Mice cloned from embryonic stem cells. Proc Natl Acad Sci USA 1999; 96(26):14984-14989.
113. Visscher PM, Smith D, Hall SJ et al. A viable herd of genetically uniform cattle. Nature 2001; 409(6818):303.
114. Eggan K, Akutsu H, Hochedlinger K et al. X-Chromosome inactivation in cloned mouse embryos. Science 2000; 290(5496):1578-1581.
115. Nolen LD, Gao S, Han Z et al. X chromosome reactivation and regulation in cloned embryos. Dev Biol 2005; 279(2):525-540.
116. Xue F, Tian XC, Du F et al. Aberrant patterns of X chromosome inactivation in bovine clones. Nat Genet 2002; 31(2):216-220.
117. Wakayama T, Tabar V, Rodriguez I et al. Differentiation of embryonic stem cell lines generated from adult somatic cells by nuclear transfer. Science 2001; 292(5517):740-743.
118. Gao S, Czirr E, Chung YG et al. Genetic variation in oocyte phenotype revealed through parthenogenesis and cloning: Correlation with differences in pronuclear epigenetic modification. Biol Reprod 2004; 70(4):1162-1170.
119. Tecirlioglu RT, Cooney MA, Lewis IM et al. Comparison of two approaches to nuclear transfer in the bovine: Hand-made cloning with modifications and the conventional nuclear transfer technique. Reprod Fertil Dev 2005; 17(5):573-585.
120. Li GP, Bunch TD, White KL et al. Development, chromosomal composition, and cell allocation of bovine cloned blastocyst derived from chemically assisted enucleation and cultured in conditioned media. Mol Reprod Dev 2004; 68(2):189-197.

121. Fulka Jr J, Loi P, Fulka H et al. Nucleus transfer in mammals: Noninvasive approaches for the preparation of cytoplasts. Trends Biotechnol 2004; 22(6):279-283.
122. Li GP, White KL, Bunch TD. Review of enucleation methods and procedures used in animal cloning: State of the art. Cloning Stem Cells 2004; 6(1):5-13.
123. Kim TM, Hwang WS, Shin JH et al. Development of a nonmechanical enucleation method using X-ray irradiation in somatic cell nuclear transfer. Fertil Steril 2004; 82(4):963-965.
124. Galli C, Lagutina I, Vassiliev I et al. Comparison of microinjection (piezo-electric) and cell fusion for nuclear transfer success with different cell types in cattle. Cloning Stem Cells 2002; 4(3):189-196.
125. Brind S, Swann K, Carroll J. Inositol 1,4,5-trisphosphate receptors are downregulated in mouse oocytes in response to sperm or adenophostin A but not to increases in intracellular Ca(2+) or egg activation. Dev Biol 2000; 223(2):251-265.
126. Jellerette T, He CL, Wu H et al. Downregulation of the inositol 1,4,5-trisphosphate receptor in mouse eggs following fertilization or parthenogenetic activation. Dev Biol 2000; 223(2):238-250.
127. Kishikawa H, Wakayama T, Yanagimachi R. Comparison of oocyte-activating agents for mouse cloning. Cloning 1999; 1(3):153-159.
128. Gao S, Chung YG, Williams JW et al. Somatic cell-like features of cloned mouse embryos prepared with cultured myoblast nuclei. Biol Reprod 2003; 69(1):48-56.
129. Thompson JG, McNaughton C, Gasparrini B et al. Effect of inhibitors and uncouplers of oxidative phosphorylation during compaction and blastulation of bovine embryos cultured in vitro. J Reprod Fertil 2000; 118(1):47-55.
130. Hill JR, Burghardt RC, Jones K et al. Evidence for placental abnormality as the major cause of mortality in first-trimester somatic cell cloned bovine fetuses. Biol Reprod 2000; 63(6):1787-1794.
131. Lee RS, Peterson AJ, Donnison MJ et al. Cloned cattle fetuses with the same nuclear genetics are more variable than contemporary half-siblings resulting from artificial insemination and exhibit fetal and placental growth deregulation even in the first trimester. Biol Reprod 2004; 70(1):1-11.
132. Wells DN, Forsyth JT, McMillan V et al. The health of somatic cell cloned cattle and their offspring. Cloning Stem Cells 2004; 6(2):101-110.
133. Everitt GC, Jury KE, Dalton DC et al. Beef production from the dairy herd. I Calving records from Friesian cows mated to Friesian and beef breed bulls. New Zealand J Agricult Res 1978; 21:197-208.
134. Morrow C, Berg M, McDonald R et al. Composition of allantoic fluid in cattle pregnant with AI-, IVP- or nuclear transfer-generated embryos. Reprod Fertil Dev 2005; 17:177.
135. Wells DN, Misica PM, Tervit HR. Production of cloned calves following nuclear transfer with cultured adult mural granulosa cells. Biol Reprod 1999; 60(4):996-1005.
136. Wells DN. Cloning in livestock agriculture. Reproduction 2003; (Supplement 61):131-150.
137. Hill JR, Roussel AJ, Cibelli JB et al. Clinical and pathologic features of cloned transgenic calves and fetuses (13 case studies). Theriogenology 1999; 51(8):1451-1465.
138. Renard JP, Zhou Q, LeBourhis D et al. Nuclear transfer technologies: Between successes and doubts. Theriogenology 2002; 57(1):203-222.
139. Heyman Y, Zhou Q, Lebourhis D et al. Novel approaches and hurdles to somatic cloning in cattle. Cloning Stem Cells 2002; 4(1):47-55.
140. Lanza RP, Cibelli JB, Faber D et al. Cloned cattle can be healthy and normal. Science 2001; 294(5548):1893-1894.
141. Pace MM, Augenstein ML, Betthauser JM et al. Ontogeny of cloned cattle to lactation. Biol Reprod 2002; 67(1):334-339.
142. Archer GS, Friend TH, Piedrahita J et al. Behavioral variation among cloned pigs. Applied Animal Behaviour Science 2003; 82(2):151-161.
143. Tamashiro KL, Wakayama T, Blanchard RJ et al. Postnatal growth and behavioral development of mice cloned from adult cumulus cells. Biol Reprod 2000; 63(1):328-334.
144. Wilson JM, Williams JD, Bondioli KR et al. Comparison of birth weight and growth characteristics of bovine calves produced by nuclear transfer (cloning), embryo transfer and natural mating. Anim Reprod Sci 1995; 38:73-83.
145. Walsh MK, Lucey JA, Govindasamy-Lucey S et al. Comparison of milk produced by cows cloned by nuclear transfer with milk from noncloned cows. Cloning and Stem Cells 2003; 5(3):213-219.
146. Ogura A, Inoue K, Ogonuki N et al. Phenotypic effects of somatic cell cloning in the mouse. Cloning Stem Cells 2002; 4(4):397-405.
147. Tamashiro KL, Wakayama T, Akutsu H et al. Cloned mice have an obese phenotype not transmitted to their offspring. Nat Med 2002; 8(3):262-267.
148. Renard JP, Chastant S, Chesne P et al. Lymphoid hypoplasia and somatic cloning. Lancet 1999; 353(9163):1489-1491.

149. Ogonuki N, Inoue K, Yamamoto Y et al. Early death of mice cloned from somatic cells. Nat Genet 2002; 30(3):253-254.
150. Carroll JA, Carter DB, Korte S et al. The acute-phase response of cloned pigs following an immune challenge. American Society Of Animal Science, Southern Section Meeting, 2004, (abstract).
151. Ohta H, Wakayama T. Generation of normal progeny by intracytoplasmic sperm injection following grafting of testicular tissue from cloned mice that died postnatally. Biol Reprod 2005; 73(3):390-395.
152. Lane N, Dean W, Erhardt S et al. Resistance of IAPs to methylation reprogramming may provide a mechanism for epigenetic inheritance in the mouse. Genesis 2003; 35(2):88-93.
153. Rakyan VK, Chong S, Champ ME et al. Transgenerational inheritance of epigenetic states at the murine Axin(Fu) allele occurs after maternal and paternal transmission. Proc Natl Acad Sci USA 2003; 100(5):2538-2543.
154. Roemer I, Reik W, Dean W et al. Epigenetic inheritance in the mouse. Curr Biol 1997; 7(4):277-280.
155. Wells DN, Misica PM, Tervit HR et al. Adult somatic cell nuclear transfer is used to preserve the last surviving cow of the Enderby Island cattle breed. Reprod Fertil Dev 1998; 10(4):369-378.
156. Woolliams JA, Wilmut I. New advances in cloning and their potential impact on genetic variation in livestock. Anim Sci 1999; 68:245-256.
157. Wells DN. The integration of cloning by nuclear transfer in the conservation of animal genetic resources. In: Simm G, Villanuva B, DSK, Townsend S, eds. Farm Animal Genetic Resources. British Society of Animal Science, 2004:30:223-241.
158. Mackle TR, Bryant AM, Petch SF et al. Nutritional influences on the composition of milk from cows of different protein phenotypes in New Zealand. J Dairy Sci 1999; 82(1):172-180.
159. Hein WR, Griebel PJ. A road less travelled: Large animal models in immunological research. Nat Rev Immunol 2003; 3(1):79-84.
160. Archer GS, Dindot S, Friend TH et al. Hierarchical phenotypic and epigenetic variation in cloned swine. Biol Reprod 2003; 69(2):430-436.
161. Lanza R, Shieh JH, Wettstein PJ et al. Long-term bovine hematopoietic engraftment with clone-derived stem cells. Cloning Stem Cells 2005.
162. Lanza RP, Chung HY, Yoo JJ et al. Generation of histocompatible tissues using nuclear transplantation. Nat Biotechnol 2002; 20(7):689-696.
163. Laible G, Wells DN. Transgenic cattle applications: the transition from promise to proof. In: Harding, SE, ed. Biotechnology & Genetic Engineering Reviews, Vol. 22. Paris: Lavoisier Publishing, 2006:125-50.
164. Schnieke AE, Kind AJ, Ritchie WA et al. Human factor IX transgenic sheep produced by transfer of nuclei from transfected fetal fibroblasts. Science 1997; 278(5346):2130-2133.
165. Kuroiwa Y, Kasinathan P, Matsushita H et al. Sequential targeting of the genes encoding immunoglobulin-mu and prion protein in cattle. Nat Genet 2004; 36(7):775-780.
166. Forsyth JT, Troskie HE, Brophy B et al. Utilising preimplantation genetic diagnosis and OPU-IVP-ET to generate multiple progeny of predetermined genotype from cloned transgenic heifers. Reproduction, Fertility and Development 2005; 17, (Abstract 330).
167. Siewerdt F, Eisen EJ, Murray JD. Direct and correlated responses to short-term selection for 8-week body weight in lines of transgenic (oMt1a-oGH) mice. In: Murray JD, Anderson GB, Oberbauer AM et al, eds. Transgenic Animals in Agriculture. Oxon, UK: CABI Publishing, 1999:231-250.
168. Brink MF, Bishop MD, Pieper FR. Developing efficient strategies for the generation of transgenic cattle which produce biopharmaceuticals in milk. Theriogenology 2000; 53(1):139-148.
169. Rudolph NS. Biopharmaceutical production in transgenic livestock. Trends Biotechnol 1999; 17(9):367-374.
170. Zhu L, van de Lavoir MC, Albanese J et al. Production of human monoclonal antibody in eggs of chimeric chickens. Nat Biotechnol 2005; 23(9):1159-1169.
171. Robl JM, Kasinathan P, Sullivan E et al. Artificial chromosome vectors and expression of complex proteins in transgenic animals. Theriogenology 2003; 59(1):107-113.
172. Karatzas CN. Designer milk from transgenic clones. Nat Biotechnol 2003; 21(2):138-139.
173. Wall RJ, Kerr DE, Bondioli KR. Transgenic dairy cattle: Genetic engineering on a large scale. J Dairy Sci 1997; 80(9):2213-2224.
174. Brophy B, Smolenski G, Wheeler T et al. Cloned transgenic cattle produce milk with higher levels of beta-casein and kappa-casein. Nat Biotechnol 2003; 21(2):157-162.
175. Muller M, Brem G. Transgenic approaches to the increase of disease resistance in farm animals. Rev Sci Tech 1998; 17(1):365-378.
176. Bueler H, Aguzzi A, Sailer A et al. Mice devoid of PrP are resistant to scrapie. Cell 1993; 73(7):1339-1347.

177. Perrier V, Kaneko K, Safar J et al. Dominant-negative inhibition of prion replication in transgenic mice. Proc Natl Acad Sci USA 2002; 99(20):13079-13084.
178. Denning C, Burl S, Ainslie A et al. Deletion of the alpha(1,3)galactosyl transferase (GGTA1) gene and the prion protein (PrP) gene in sheep. Nat Biotechnol 2001; 19(6):559-562.
179. Cyranoski D. Koreans rustle up madness-resistant cows. 2003; 426(6968):743.
180. Wall RJ, Powell AM, Paape MJ et al. Genetically enhanced cows resist intramammary Staphylococcus aureus infection. Nat Biotechnol 2005; 23(4):445-451.
181. Ward KA, Brownlee AG, Leish Z et al. Proceedings of the VII world conference on animal production. Edmonton, Canada: 1993:1:267.
182. Ward KA. Transgene-mediated modifications to animal biochemistry. Trends Biotechnol 2000; 18(3):99-102.
183. Golovan SP, Meidinger RG, Ajakaiye A et al. Pigs expressing salivary phytase produce low-phosphorus manure. Nat Biotechnol 2001; 19(8):741-745.
184. Forsberg EJ. Commercial applications of nuclear transfer cloning: Three examples. Reprod Fertil Dev 2005; 17(2):59-68.
185. Tsunoda Y, Kato Y. Full-term development after transfer of nuclei from 4-cell and compacted morula stage embryos to enucleated oocytes in the mouse. J Exp Zool 1997; 278(4):250-254.
186. Humpherys D, Eggan K, Akutsu H et al. Epigenetic instability in ES cells and cloned mice. Science 2001; 293(5527):95-97.
187. Yabuuchi A, Yasuda Y, Kato Y et al. Effects of nuclear transfer procedures on ES cell cloning efficiency in the mouse. J Reprod Dev 2004; 50(2):263-268.
188. Ogura A, Inoue K, Takano K et al. Birth of mice after nuclear transfer by electrofusion using tail tip cells. Mol Reprod Dev 2000; 57(1):55-59.
189. Wakayama T, Yanagimachi R. Cloning of male mice from adult tail-tip cells. Nat Genet 1999; 22(2):127-128.
190. Ogura A, Inoue K, Ogonuki N et al. Production of male cloned mice from fresh, cultured, and cryopreserved immature Sertoli cells. Biol Reprod 2000; 62(6):1579-1584.
191. Kato Y, Tani T, Tsunoda Y. Cloning of calves from various somatic cell types of male and female adult, newborn and fetal cows. J Reprod Fertil 2000; 120(2):231-237.
192. Galli C, Duchi R, Moor RM et al. Mammalian leukocytes contain all the genetic information necessary for the development of a new individual. Cloning 1999; 1(3):161-170.

CHAPTER 4

Centrosome Inheritance after Fertilization and Nuclear Transfer in Mammals

Qing-Yuan Sun and Heide Schatten*

Abstract

Centrosomes, the main microtubule organizing centers in a cell, are nonmembrane-bound semi-conservative organelles consisting of numerous centrosome proteins that typically surround a pair of perpendicularly oriented cylindrical centrioles. Centrosome matrix is therefore oftentimes referred to as pericentriolar material (PCM). Through their microtubule organizing functions centrosomes are also crucial for transport and distribution of cell organelles such as mitochondria and macromolecular complexes. Centrosomes undergo cell cycle-specific reorganizations and dynamics. Many of the centrosome-associated proteins are transient and cell cycle-specific while others, such as γ-tubulin, are permanently associated with centrosome structure. During gametogenesis, the spermatozoon retains its proximal centriole while losing most of the PCM, whereas the oocyte degenerates centrioles while retaining centrosomal proteins. In most mammals including humans, the spermatozoon contributes the proximal centriole during fertilization. Biparental centrosome contributions to the zygote are typical for most species with some exceptions such as the mouse in which centrosomes are maternally inherited and centrioles are assembled de novo during the blastocyst stage. After nuclear transfer in reconstructed embryos, the donor cell centrosome complex is responsible for carrying out functions that are typically fulfilled by the sperm centrosome complex during normal fertilization, including spindle organization, cell cycle progression and development. In rodents, donor cell centrioles are degraded after nuclear transfer, and centrosomal proteins from both donor cell and recipient oocytes contribute to mitotic spindle assembly. However, questions remain about the faithful reprogramming of centrosomes in cloned mammals and its consequences for embryo development. The molecular dynamics of donor cell centrosomes in nuclear transfer eggs need further analysis. The fate and functions of centrosome components in nuclear transfer embryos are being investigated by using molecular imaging of centrosome proteins labeled with specific markers including, but not limited to, green fluorescent protein (GFP).

Introduction to the Centrosome

The centrosome, approximately $1\mu m$ in size, is a nonmembrane-bound, semi-conservative organelle with a highly complex molecular composition and dynamic architecture in animal cells. The centrosome is generally known as the cell's main microtubule organizing center that is closely associated with the nucleus in interphase; it undergoes cell cycle-specific reorganizations and forms the center of the mitotic poles during mitosis. Each mammalian somatic cell

*Corresponding Author: Heide Schatten—Department of Veterinary Pathobiology, University of Missouri-Columbia, 1600 E. Rollins Street, Columbia, Missouri 65211, U.S.A. Email: SchattenH@Missouri.Edu

Somatic Cell Nuclear Transfer, edited by Peter Sutovsky. ©2007 Landes Bioscience and Springer Science+Business Media.

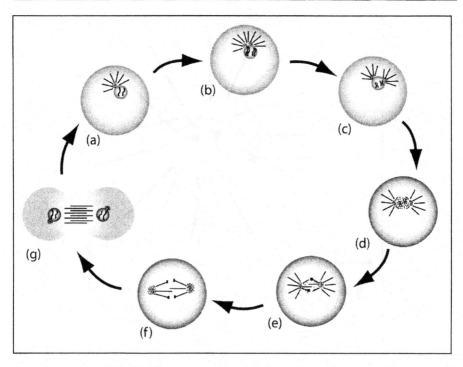

Figure 1. Centrosome duplication, centrosome separation and spindle organization in a mammalian cell cycle. Interphase cells contain a single centrosome from which interphase microtubules are nucleated (a). During S-phase, the centrosome duplicates concomitantly with DNA replication (b). During prophase, the duplicated centrosomes migrate to the opposite sides of the nucleus (c). After nuclear envelope breakdown, elongating astral microtubules nucleate from the bipolar centrosomes and attach to the kinetochores of the condensed chromosomes (d). At metaphase, each chromosome is captured by microtubules originating from both spindle poles, and all chromosomes are aligned at the spindle equator (e). After chromosome segregation in anaphase (f) and telophase (g), two new daughter cells are produced, each with one centrosome associated with the nucleus. Modified from Varmark H (2004).[4]

typically contains one centrosome, which is precisely reproduced in coordination with DNA replication at the S phase during interphase. After duplication the two separated centrosomes move to opposite sides of the nucleus and undergo a complex process of maturation that enables them to function as mitotic centrosomes. During cell division, centrosomes form the poles of the bipolar mitotic spindle. After anaphase and telophase, centrosomes become separated into two daughter cells (Fig. 1). The main function of centrosomes is the organization of both interphase microtubule arrays responsible for cell polarity and the mitotic spindle, which mediates the strictly bipolar separation of chromosomes. In addition to these established functions, the centrosome is also implicated in numerous other cellular functions and impacts cellular processes such as entry into mitosis, cytokinesis, G1/S transition, and monitoring of DNA damage.[1] Aberrant duplication of centrosomes has been implicated in many kinds of tumors in which multiple centrosomes organize multipolar spindles, resulting in abnormal chromosome segregation and aneuploidy.[2] Structural defects of centrosomes can also account for the formation of abnormal mitosis and multipolar cells frequently observed in cancer.[3]

A typical centrosome consists of a pair of perpendicularly oriented cylinder-shaped centrioles surrounded by an electron dense protein matrix referred to as pericentriolar material (PCM Fig. 2). The wall of the centriole cylinder is composed of nine parallel bundles of microtubule

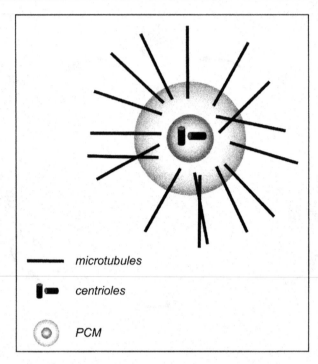

microtubules

centrioles

PCM

Figure 2. Fine structure of a centrosome. The centrosome is composed of two centrioles surrounded by a protein matrix called the pericentriolar material (PCM). The PCM is a complex meshwork of proteins, including γ-tubulin containing ring complex (γ-TuRC) that nucleate microtubules. Modified from Varmark H (2004).[4]

triplets, while the PCM consists of a complex meshwork of proteins, including γ-tubulin containing ring complexes (γ-TuRC) that nucleate the polymerization of microtubules and proteins that regulate centrosome duplication and functions.[4]

γ-Tubulin and centrin are two classical markers of the centrosome with specific localizations, but the majority of these two proteins (80-90%) are not centrosome-associated. Matrix proteins are concentrated at the centrosome, with specific localizations while the cytoplasmic pool of matrix proteins is small. The localizations of many centrosome proteins such as γ-tubulin, centrin, ninein, and pericentrin have been reported.[5,6] It is hypothesized that the microtubule-minus-end-binding proteins, including γ-TuRC, are concentrated at the proximal end of centrioles, whereas tubulin polyglutamylation of the centriole walls modulates interaction between tubulin and microtubule associated proteins. These centriole-interacting proteins can bind and accumulate matrix proteins. The matrix is in turn capable of binding and concentrating γ-TuRC, as well as carrying out regulatory activities. Pericentrin anchors γ-tubulin complexes at centrosomes in mitotic cells, which is required for proper spindle organization.[7] One of the essential cell cycle-dependent centrosome-associated proteins is the nuclear mitotic apparatus protein (NuMA) that is distributed to the separating centrosomes during early mitosis, ensures minus-end binding and stabilization on the centrosome side facing the chromosomes, resulting in the cross-linking of spindle microtubules, which is essential for the organization and stabilization of spindle poles from early mitosis until at least the onset of anaphase.[5,8]

Although most animal cell divisions occur in the presence of centrosomes, it is now well established that plant cells and specialized animal cells such as eggs (described in detail below) can assemble bipolar spindles in the absence of typical centrosomes,[9] and that somatic cells can also use a centrosome-independent pathway for spindle formation that is normally repressed in the presence of the centrosome.[10]

To understand embryos it is important to understand centrosome dynamics and molecular restructuring during gametogenesis and fertilization.

Centrosome Reduction during Gametogenesis

Animal spermatids and primary oocytes initially have typical centrosomes, showing the hallmark "9+0" organization of microtubule triplets surrounded by pericentriolar material, common for somatic cells. These somatic cell-like centrosomes undergo degeneration and profound modification during the final stages of gametogenesis to meet the specific needs of gamete functions and fertilization (Fig. 3). It has been shown that during gametogenesis, spermatozoa retain centrioles but lose most of the pericentriolar centrosomal proteins. In a complementary fashion, the oocytes lose centrioles while retaining a stockpile of centrosomal proteins (for review see refs. 11,12). It is hypothesized that the reciprocal reduction of centrosomal constituents in the spermatozoon and the oocyte shape centrosomal material to be complementary to each other after fertilization, forming a biparental centrosome in the zygote.[13]

During centrosome reduction in spermatogenesis, centrosomes first lose microtubule-nucleating function, then lose centrosomal proteins, and finally lose centrioles. In the mouse, the round spermatids display normal centrosomes consisting of a pair of centrioles along with γ-tubulin- and centrin-containing foci. However, they do not seem to organize microtubules. Elongating spermatids display γ-tubulin and centrin spots in the connecting piece region at the junction of the nascent sperm head and flagellum, while microtubules are transiently organized from the perinuclear ring as the manchette. Electron microscopy studies using immunogold labeling revealed that γ-tubulin and centrin are mainly localized in the centriolar adjunct from which an aster of microtubules emanates. γ-tubulin and centrin diminish from the neck region and are discarded in the residual bodies during spermiation. The distal centriole degenerates during the testicular stage of spermiogenesis, while the proximal centriole is lost during the epididymal stage.[14,15] Complete centriolar degeneration has also been reported in rat spermatozoa.[16]

In rhesus monkeys and humans, centrosomes are reduced during spermiogenesis, but not as completely as in mice. All γ-tubulin and half of the centrin are discarded, but the remaining portion is still bound to the centriole microtubules in fully differentiated rhesus monkey spermatozoa. Rhesus and human spermatozoa have only proximal centrioles intact, whereas the distal centrioles are mostly disorganized or highly degenerated.[17,18] The proximal centrioles remain intact but the distal centrioles undergo various degrees of degeneration in several other species.[11]

Contrary to spermatogenesis, proteins that are essential for microtubule nucleation, like γ-tubulin, are retained in the cytoplasm during oogenesis while the centrioles degenerate. In the mouse, oogonia and fetal oocytes display normal centrioles until pachytene stage. Centrioles disappear from the oocyte during early oogenesis, some time after the pachytene stage of prophase I of meiosis at about the time oocytes enter a resting (dictyate) stage, and they are not detected until the blastocyst stage.[19,20] Instead, multiple structures known as microtubule organizing centers (MTOC) form the spindle poles during meiosis. MTOCs are immunologically similar to pericentriolar material, containing γ-tubulin, pericentrin, NuMA etc.[19,21-23] Centrosomes display precisely controlled spatio-temporal changes during the onset of meiotic maturation. Accumulation of centrosomal proteins to a single locus followed by a sequestration to several spots occurs during nuclear envelope breakdown.[24] It has been reported that before the onset of oocyte maturation, two large γ-tubulin-positive structures, averaging 10 μm in diameter, are found in association with the oocyte cortex. These structures called multivesicular aggregates (MVA) contain γ-tubulin, but they do not resemble MTOCs morphologically nor do they appear to nucleate MTs. At the onset of maturation, these two MVA migrate toward the GV breaking into smaller units, and only some of them mature into MTOCs and nucleate MTs.[19] During the germinal vesicle breakdown (GVBD) stage, multiple perinuclear MTOCs appear. During the first and second meioses, acentriolar poles of meiotic spindles are organized exclusively by MTOC material. In addition, there are numerous microtubule asters organized by MTOCs in the cytoplasm of mature oocytes. It has been reported that at metaphase of meiosis I and II the microtubule nucleation capacity of centrosomes is diminished in mouse oocytes.[25] Furthermore, pericentrin incorporation into spindles is not required for meiotic spindle formation, but its dynamic reorganization may be involved in critical cell cycle transitions during mouse oocyte meiotic maturation.[26]

The absence of centrioles in the poles of MII meiotic spindles has also been reported in rabbits, cows, sheep, and humans.[11,27,28] The pericentriolar centrosomal proteins are distributed as concentric poles of the barrel-shaped spindles during dividing stages. In humans, the first MTOCs are perinuclear, but the number and distribution increases widely as the oocytes enter metaphase.[29] However, no cytoplasmic MTOCs exist in mature oocytes. Human oocytes have no granular centrosomal material at meiotic spindle poles, in contrast to mouse oocytes which have dominant maternal centrosomes.[30]

Centrosome Inheritance after Fertilization

Fertilization and early development requires the control of maternal and paternal centrosome components to ensure that cells never contain more than two functional centrosomes. If both female and male gametes contributed centrosomes that are capable of replication and function during fertilization, assembly of tetrapolar spindles and abortive development would ensue. Although it has long been recognized that both the sperm and the egg contribute equal haploid genomes during fertilization, the relative contributions of the centrosome by each gamete have only been documented recently. It is now clear that in the mouse, there is no evidence of a functional centrosome in the sperm, and centrosomes are maternally inherited, while in humans and other mammals, the spermatozoa contribute the proximal centriole during fertilization. Recent observations support a biparental centrosomal contribution to zygotes in almost all mammals examined except for rodents, in contrast to a strictly paternal inheritance pattern that had been suggested in earlier studies.

In most mammals including humans, the oocyte centrosome is greatly reduced and inactivated. MII oocytes lack centrioles, although functional MTOCs are present at their spindle poles. However, a functional centrosomal structure is restored and duplicated after fertilization in the zygote. In human fertilization, the sperm tail and its centriole-harboring connecting piece are incorporated into the ooplasm together with the sperm head. Most sperm cytoplasmic structures including mitochondria, fibrous sheath, microtubule doublets, outer dense fibers, and the striated columns of the connecting piece, are discarded in a programmed order.[31] The proximal centriole remains intact and forms the sperm aster while the sperm head is decondensing in the ooplasm. The sperm aster ensures pronuclear apposition, leading to close congregation of the male and female chromosomes into a single mitotic metaphase plate and a bipolar spindle formation. The centriole duplicates during the pronuclear stage, and at syngamy, centrioles are found at opposite poles during first cleavage. Typical centrioles showing the characteristic "9+0" triplets of microtubules are evident. During centriolar replication, the daughter centriole grows laterally from the parent and gradually acquires pericentriolar material (PCM). The two centrioles are surrounded by a halo of electron-dense PCM that nucleates microtubules, thus constituting a typical centrosome.[31] Centrioles are detected at all stages of embryonic cleavage from the 1-cell through 8-cell stages, right up to the hatching blastocyst stage. Thus, the zygote centrosomes are ancestors of centrosomes in embryonic, fetal and adult somatic cells.[32,33] The sperm centrosome has important implications in human infertility, and it is postulated that sperm centrosomal dysfunction could lead to aberrant embryonic development based on centriolar defects in sperm with impaired motility.[30,33] This has been proven in a globozoospermic patient, though an attempt to restore defective human sperm centrosomal function has been performed, with no success.[34,35]

Although centrioles are paternally inherited in humans, formation of typical functional centrosomes in fertilized eggs requires blended contributions of the paternal centriole and maternal pericentriolar constituents. The latter is attracted to a paternally- introduced template to form functional zygote centrosomes after fertilization. Some centrosomal proteins such as γ-tubulin are biparentally inherited in humans and perhaps other species. Recruitment of maternal γ-tubulin to the sperm's centrosomal γ-tubulin results in significant increases of PCM content after sperm exposure to the ooplasmic environment, particularly after sperm "priming" induced by disulfide bond reduction. Sperm also contain centrin, which is thought to be important for the reorganization of the sperm centrosomal complex after insemination and perhaps for subsequent splitting of the early zygotic centrosome.[36] The centrosome inheritance of nonhuman primates during fertilization is similar to that of humans.[37]

Electron microscopy studies revealed that there are no centrioles in mature sheep oocytes, but a small microtubule aster is formed in the sperm neck region containing the centriole after fertilization. In addition, one or two centrioles are observed in/or close to the sperm connecting piece region lying between the apposed pronuclei.[38] Centrioles were not detected in the mitotic spindle of sheep parthenogenotes, while they were found to exist at the opposite poles of the first mitotic spindle in monospermic eggs and androgenetic eggs, suggesting that centrioles of sheep zygotes are paternally derived.[27] In rabbits, centrioles are not observed in early stage parthenogenetic embryos until the blastocyst stages, when centrioles develop de novo in blastocysts.[39] During fertilization of bovine oocytes, maternal γ-tubulin is recruited by sperm components to reconstitute the zygotic centrosome. In the absence of sperm components, oocyte γ-tubulin nucleates and organizes microtubules to position the female pronucleus and spindle during the first cell cycle of bovine parthenotes.[40] However, since bovine sperm also contain pericentriolar material, it is assumed that both paternal centrosomes (centrioles) and maternal centrosomes (pericentriolar material) are involved in the organization of bipolar spindles in early embryos.[36,41] Biparental centrosome contribution is also the case in sheep, pig, rabbit and horse zygotes, in which functional centrosomes form during fertilization as a result of blending of paternal and maternal centrosome components.[27,42-44] It is widely accepted that zygotic centrosomes are biparentally contributed in mammals except for rodents.

The most common strategy of centrosome formation in mammalian fertilized eggs involves differential contributions of centrioles and pericentriolar material (PCM) from each gamete. Rodents have a unique centrosome inheritance strategy, i.e., maternal centrosome inheritance. The mouse oocyte does not tolerate centrioles. In early embryos before the blastocyst stage typical centrosomes containing centrioles are absent; while in 4-4.5 day-old blastocysts centrosomes with double centrioles form de novo.[45,46]

In unfertilized mouse oocytes, the meiotic spindle poles are displayed as broad-beaded centrosomes. In addition, centrosomal material is detected in the cytoplasm as aggregated clusters, about 16 in number, which are foci for small aster-like arrays of microtubules. Centrosomal proteins like protein 5051, pericentrin, and γ-tubulin are localized to the astral foci.[11] After sperm incorporation, as the pronuclei develop and more cytoplasmic microtubules assemble, a few of the foci associate with the peripheries of the nuclei. The number of foci multiplies during the first cell cycle.[47] Fertilizing sperm may also contain γ-tubulin since it has been shown at the cytastral centers as well as in the incorporated sperm basal body complex after fertilization, and the γ-tubulin foci coalesce at the perinuclear microtubule organizing regions of the two pronuclei at the first mitotic prophase.[13] During mitosis, γ-tubulin is found in association with broad bands that form the poles of the first mitotic spindle. By the late preimplantation stage, when newly generated centrioles have been reported to arise, γ-tubulin remains localized at the centrosome of mitotic cells.[21,48] By studying parthenogenetic and polyspermic mouse oocytes, it was also shown that centrosomes are maternally inherited in fertilized mouse eggs.[49]

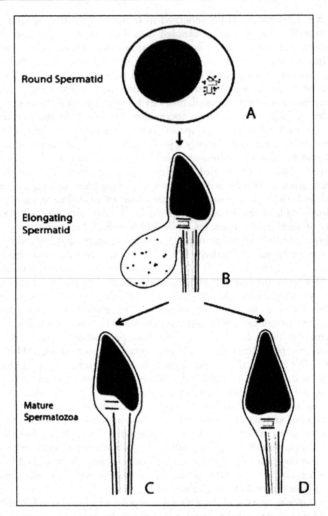

Figure 3. Centrosome reduction during spermiogenesis. Round spermatids possess an intact centrosome containing a pair of centrioles and centrosomal proteins (A). During spermiogenesis, the microtubules of the distal centriole extend as axoneme, and a large part of the centrosomal proteins is discarded with the residual bodies (B). Rodent spermatozoa lose both centrioles completely (C), whereas other mammalian spermatozoa retain proximal centrioles (D). Modified from Manandhar G et al, 2005.[11]

Centrosome Inheritance after Nuclear Transfer

As detailed above, the studies on gametogenesis and fertilization have provided insights into the complex molecular regulation and functions of centrosomes in reproductive cells, but complexities are added when somatic cell centrosomes are challenged to carry out functions within the enucleated oocyte that are typically carried out by tightly controlled interactions between sperm and egg centrosomal material.

Cloning of mammalian embryos is typically performed by introducing a somatic cell nucleus with known genetic value into an enucleated oocyte from a different animal (SCNT, somatic cell nuclear transfer) followed by activation of the reconstructed oocyte. The somatic cell's centrosome is tightly associated with the nucleus and is transferred along with the donor cell nucleus while the oocyte's major centrosome material localized to the meiotic spindle is removed during the enucleation process (Fig. 4). Centrosomal material in the recipient ooplasm and the centrosome of the donor cell coexist in the reconstituted embryo. So far, little is understood about the fate and role of the donor cell centrosome and remaining oocyte centrosomal material that is needed for spindle organization of reconstructed embryos. The analysis of centrosomes is crucial, as centrosomes play central roles in cell divisions, establishment of cell symmetry and asymmetry, metabolic activities, and other vital cell functions (reviewed by Schatten et al ref. 50) that will be addressed below. Cell and molecular studies are needed to determine strategies to improve the currently low cloning efficiency rate which averages ca.1-3% of SCNT constructs in sheep, mice, pigs, cattle (both domestic and wild/endangered), goats, rabbits, cats, dogs, mules, and horses) to produce live offspring.[51,52]

Somatic cell centrosomes need to be remodeled by the oocyte to assume critical functions during embryonic cell cycles and throughout the life of the developing embryo. Somatic cell cycles are different from embryonic cell cycles and employ different regulatory systems. The oocyte's regulatory systems have to communicate with the donor cell nuclear and centrosome material for embryo survival. Many of the numerous centrosome proteins (some of which are of nuclear origin) depend on nucleo-cytoplasmic interactions to carry out cell cycle-specific and developmentally regulated functions. Functional centrosomes play a central role in directing and governing metabolic activities along microtubules which include translocation of mitochondria to their specific destinations for ATP supply, translocation of macromolecular complexes, trafficking of vesicles that may contain important enzymes, and cell cycle-dependent, transiently centrosome-associated proteins that are needed for molecular centrosome restructuring to carry out cell cycle-specific centrosome functions. Subtle abnormalities can lead to developmental failures. The regulatory systems for accurate centrosome functions may need to be reprogrammed for successful cloning. Modifications in culture conditions may be needed to support the different requirements for reproductive and somatic cell systems including changes in pH and calcium that are critical for proper centrosome functions. During fertilization, the sperm triggers a program of egg activation that includes changes in pH and calcium that are only partly mimicked in parthenogenetically activated eggs[53-55] and may not be fully represented in SCNT eggs that are electrically activated. The proper pH and calcium levels are crucial for cytoskeletal regulation. We do not yet know whether pH and Ca^{++} requirements in SCNT embryos are similar to those in fertilized eggs and whether cytoskeletal organization resembles that in fertilized eggs. Other cell cycle regulators which may affect centrosome maturation and normal development include cyclin B, a major regulator of centrosomes that is crucial for centrosome maturation and cell cycle-dependent molecular centrosome dynamics.[56-58] We do not yet know how cyclin B is regulated in NT embryos. In embryonic cells, cyclin B synthesis is constant throughout the cell cycle and cyclin B accumulation is the result of decrease in its degradation rate.[50] In somatic cells, cyclin B synthesis increases during G2 and M, predominantly as a result of increase in cyclin B gene transcription. Subtle differences in regulatory mechanisms may affect cloning success.

Centrosomes have been studied in various cell systems including somatic cells in which specific centrosome proteins and their cell-cycle dependent regulation have been addressed. Centrosome abnormalities are associated with various diseases including cancer which has been related to imbalances in phosphorylation and centrosome protein compositions. These studies emphasize the importance of precise centrosome regulation. In cloned embryos, a few studies have addressed centrosome organization after nuclear transfer with focus on specific centrosome proteins but the regulation of centrosomes and signal transduction in NT embryos has not yet been addressed. Shin et al (2002)[59] used bovine SCNT embryos and observed a microtubule aster containing a γ-tubulin spot near the transferred nucleus in most reconstructed eggs regardless of activation conditions. Premature chromosome condensation (PCC) of the transferred fibroblast nuclei in nonactivated oocytes that divided into two chromosome clusters was associated with microtubule formations surrounding condensed chromosomes, and γ-tubulin

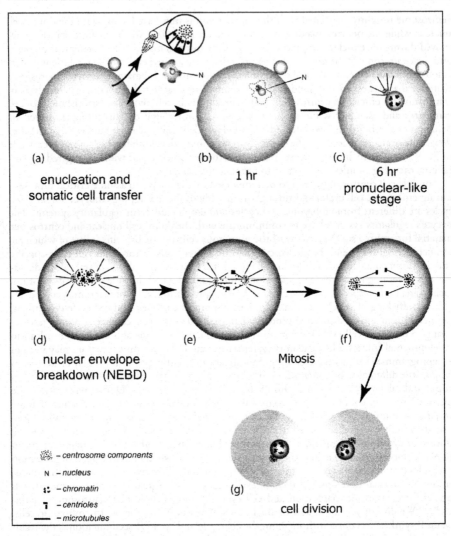

Figure 4. Centrosome organization after nuclear transfer. Cloning of mammalian embryos is performed by removing the oocyte's meiotic spindle containing DNA, microtubules, and the oocyte's major centrosomal material which is followed by fusion or injection of a somatic donor cell containing the nucleus (N) and associated centrosome complex (a). At 1 hr after nuclear transfer, the centrosome and nuclear material decompact (b). Donor cell centrosomes organize microtubules into an aster (c); centrosome material separates and moves to two opposite poles while organizing a bipolar mitotic spindle (d) that separates and moves chromosomes to the poles seen here at prometaphase (e) and anaphase (f) followed by cell division and relocalization of centrosomes that are closely associated with the nucleus of the two dividing daughter cells (g). Centrosome separation and duplication follows a typical mammalian cell cycle as described in (Fig. 1).

was detected in association with the condensed chromosomes. Two pronuclear-like structures near the microtubular aster containing γ-tubulin spot(s) later formed a syngamy-like nuclear structure.[59] In pig SCNT, similar observations have been reported.[60] These results indicate that the donor centrosome introduced during SCNT, like the centrosome introduced during fertilization, assembles microtubules for cell division and subsequent development of reconstructed embryos. These studies reported observations on the centrosome-intrinsic protein γ-tubulin up to two-cell stages. Further studies are needed to address centrosome abnormalities and to trace a large number of reconstructed oocytes and follow their individual centrosome fates, perhaps in combination with live observations using fluorescent tracers.

As mentioned above, nuclear and centrosome cycles are typically well coordinated to ensure proper segregation and distribution of the genome to the dividing daughter cells. The intimate relationship between nuclear and centrosome proteins is further apparent by the fact that many centrosome proteins are of nuclear origin which couples nuclear remodeling to centrosome remodeling. One of the best-studied nuclear-centrosome proteins is the nuclear mitotic apparatus protein (NuMA) (for recent review, please see ref. Sun and Schatten, 2006).[51] Most of our knowledge of NuMA comes from studies in somatic cells. NuMA is a multifunctional protein that plays important roles in DNA replication, reorganization, and transcription during interphase (reviewed by Zeng, 2000; Sun and Schatten, 2006)[8,61] as well as in apoptosis. During mitosis and cell division, NuMA is involved in the organization of microtubules into the mitotic apparatus. In unfertilized mature oocytes NuMA is concentrated at the poles of the meiotic spindle where its functions include the organization of microtubules into the meiotic apparatus.[23] The NuMA-containing meiotic spindle is removed during the enucleation process which leaves the oocyte without evident NuMA but in reconstructed embryos the donor cell nucleus contains NuMA and provides the source for NuMA. This has recently been shown by Zhong et al[62] who used mouse MII oocytes as recipients, mouse fibroblast, rat fibroblast, or porcine granulosa cells as donor cells to reconstruct intraspecies/interspecies nuclear transfer embryos. The NuMA antibody used in these experiments did not recognize NuMA protein of mouse oocytes but recognized porcine NuMA of the granulosa cells used as donor material, allowing tracing of NuMA, leading to the conclusion that it is contributed by the donor nucleus. These studies clearly showed that the donor cell's NuMA was translocated out of the nucleus into the cytoplasm to play a role in spindle pole formation during first mitotic metaphase. Microinjection of NuMA antibody resulted in spindle disorganization and formation of multipolar spindles. These studies provide evidence that NuMA from donor cells contributes to mitotic spindle function in reconstructed intraspecies/interspecies nuclear transfer embryos in which mouse oocytes are the recipients. In reconstructed pig oocytes, Liu et al (2005)[63] also showed that NuMA is contributed by the donor cell nucleus. However, it takes ca. 6 hours for NuMA to display characteristic staining patterns in the transfer nuclei which indicates that cross-talk between the cytoplast and donor nucleus is necessary. Varied results were obtained in nonhuman primates. It was reported that during nuclear transfer, meiotic spindle removal depletes NuMA and NuMA is not detected after nuclear transfer in the resulting abnormal mitotic spindles.[64] However, these studies have not addressed the possible contributions by the donor cell nucleus and regulatory factors that may play a role in accurate NuMA distribution and functions. These studies also reported that HSET, a mitotic kinesin motor found during mitosis, is not detected in NT spindles. In a subsequent study, by using new methods, the authors reported ineffective targeting of NuMA to spindle poles in nonhuman primate nuclear transfer embryos.[51] Development was improved, but embryo transfers of 135 nuclear transfer embryos into 25 staged surrogates still did not result in convincing evidence of pregnancies.[51] Analysis of microtubule patterns in nuclear transfer embryos revealed that key centrosomal and molecular motor proteins were deficient in mitotic spindles after nuclear transfer. NuMA protein is either not detected or at barely detectable concentrations on the NT mitotic spindle. The residual NuMA, retained within the oocyte's cytoplasm following spindle extrusion and imported into the interphase nucleus following nuclear transfer, is not effectively targeted to the spindle poles. According to these investigators, the somatic cell centrosome ineffectively assembles cytoplasmic microtubules near the incorporated somatic nucleus and the mechanism recruiting NuMA to the microtubule minus-ends is interfered with in nonhuman primate nuclear transfer eggs.[51] It is possible that nonhuman primate oocytes are not able to remodel NuMA and perhaps other critical proteins in the nucleus under the current oocyte retrieval, maturation and embryo culture conditions. However, rhesus birth after embryonic cell nuclear transfer (ECNT) has been reported[65] and it is likely that improved culture conditions will provide the necessary regulatory factors that allow proper centrosome functioning in nonhuman primates (for recent reviews on culture and cloning of nonhuman primates, please see refs. 66,67).

We do not yet know how NuMA and other centrosome proteins are remodeled in cloned embryos. Studies on tissue culture cells ascribe a significant role to cdc2/cyclin B complex in translocating NuMA from the nucleus into the cytoplasm and from the cytoplasm to the mitotic poles where it forms an insoluble crescent around the centrosomal core material.[56,57] The solubility of NuMA is regulated by phosphorylation. Maintaining NuMA in the soluble state requires active phosphorylation by protein kinases while phosphatase activity is thought to convert NuMA into large insoluble complexes.[56] These processes require precise regulation by the oocyte's cytoplasm. Inactivation of cdc2/cyclin B kinase is critical for dissociation of NuMA from the mitotic centrosomes and relocation to the reforming nuclei. Failure in inactivation of cdc2/cyclin B may result in failure of NuMA to relocate to the nucleus. Instead, NuMA may relocate to the cytoplasm and form small cytoplasmic asters that may cause cell fragmentation.

Another well-studied intrinsic centrosome protein is centrin, a member of Ca^{++}-binding family of proteins that is commonly associated with centrioles and plays a role in centrosome duplication.[68] The contribution of centrin to the centrosome in SCNT embryos has recently been addressed in two studies. Zhong et al[62] studied centrin in enucleated mouse oocytes in which mouse fibroblast, rat fibroblast, or porcine granulosa cells were used as donor cells to reconstruct intraspecies/interspecies nuclear transfer embryos. In all cases, centrin was observed in donor cells before nuclear transfer but was not observed at mitotic poles of reconstructed mouse oocytes. Manandhar et al (2005)[69] studied centrin in SCNT pig embryos and found that, like during in vitro fertilization, centrin is degraded during the early stages after nuclear transfer and first appeared in the spindle poles of dividing stage blastocyst cells.

Studies on mitochondria translocation also indicate a central role for centrosomes in developmental potential of embryos. Centrosomes are crucial for translocation of mitochondria and misguided or mismatched mitochondria can result in developmental abnormalities (for review please see Schatten et al 2005; please also see Chapter 8 by Stefan Hiendleder).[50,70]

Conclusions and Future Studies

During mammalian gametogenesis, spermatozoa retain centrioles but lose most of the pericentriolar centrosomal proteins, whereas the oocytes lose centrioles while retaining centrosomal proteins. One centrosome is reconstituted from two gametes at fertilization and this involves differential contributions of centrioles and pericentriolar material from each gamete. Although we know that both centriolar and pericentriolar proteins contribute to centrosome assembly, the related mechanisms remain poorly understood at a molecular level. Centrosome inheritance is exceptional in rodents where centrioles are lost during spermatogenesis and centrosomes are maternally inherited. In this case, centrioles can form de novo from no apparent template in blastocysts, perhaps relying on a preexisting seed that resides within PCM but is not detectable with our currently available methods.[71]

After nuclear transfer, the somatic donor cell contributes a centrosome (if transferred at G0 or G1 stage) or a pair of centrosomes (if transferred at G2 or M stage). Like the centrosome introduced during fertilization, the donor cell centrosome plays an important role in microtubule assembly and subsequent development of nuclear transfer cloned embryos. However, it appears that different regulatory systems may be employed in different species. In nonhuman primate nuclear transfer eggs, the somatic cell centrosome and the remaining centrosomal proteins in oocytes ineffectively assemble cytoplasmic microtubules near the incorporated somatic cell nucleus, which may account for the difficulty of primate cloning. In rodents, donor cell centrioles may be degraded after nuclear transfer, and centrosomal proteins from both donor cell and recipient oocyte contribute to mitotic spindle assembly. All these speculations are derived from preliminary observations, and need further experimentation. Recently, Higginbotham et al[72] described the generation and characterization of a transgenic mouse line that constitutively expresses green fluorescent protein-labeled with centrin-2 (GFP-CENT2). The phenotype of the mouse is indistinguishable from wild-type, and it displays a single pair of fluorescent centrioles in cells of every organ and time point examined. GFP-centrin and GFP-γ-tubulin are also expressed and selectively incorporated into the structure of both centrioles of human breast cancer cells and PtK2 epithelial cells, respectively, making the centrosome (centriole) clearly visible in living cells.[73,74] This model will undoubtedly be helpful for clarifying the centrosome behavior and fate in nuclear transfer eggs and embryos without interrupting their viability.

Acknowledgements

We would like to thank Dr. Peter Sutovsky for inviting this paper and for critically reading the manuscript. We thank Don Connor for help with the illustrations. Part of this project was supported by NIH grant R03-HD43829-02 to HS, and grants from NSFC (30225010, 30430530,30371037) to QYS.

References

1. Kramer A, Lukas J, Bartek J. Checking out the centrosome. Cell Cycle 2004; 3(11):1390-1393.
2. Wang Q, Hirohashi Y, Furuuchi K et al. The centrosome in normal and transformed cells. DNA Cell Biol 2004; 23:475-489.
3. Schatten H, Hueser CN, Chakrabarti A. From fertilization to cancer: The role of centrosomes in the union and separation of genomic material. Microsc Res Tech 2000; 49(5):420-427.
4. Varmark H. Functional roles of centrosome in spindle assembly and organization. J Cell Biochem 2004; 91:904-914.
5. Bornens M. Centrosome composition and microtubule anchoring mechanisms. Curr Opion Cell Biol 2002; 14:25-34.
6. Jurczyk A, Gromley A, Redick S et al. Pericentrin forms a complex with intraflagellar transport proteins and polycystin-2 and is required for primary cilia assembly. J Cell Biol 2004; 166(5):637-643.
7. Zimmerman WC, Sillibourne J, Rosa J et al. Mitosis-specific anchoring of gamma tubulin complexes by pericentrin controls spindle organization and mitotic entry. Mol Biol Cell 2004; 15(8):3642-3657.
8. Zeng C. NuMA: A nuclear protein involved in mitotic centrosome function. Microsc Res Tech 2000; 49(5):467-477.
9. Meraldi P, Nigg EA. The centrosome cycle. FEBS Lett 2002; 521:9-13.
10. Khodjakov A, Cole RW, Oakley BR et al. Centrosome-independent mitotic spindle formation in vertebrates. Curr Biol 2002; 10:59-67.
11. Manandhar G, Schatten H, Sutovsky P. Centrosome reduction during gametogenesis and its significance. Biol Reprod 2005; 72:2-13.
12. Manandhar G, Simerly C, Schatten G. Centrosome reduction during mammalian spermiogenesis. Curr Top Dev Biol 2000; 49:343-363.
13. Schatten G. The centrosome and its mode of inheritance: The reduction of centrosome during gametogenesis and its restoration during fertilization. Dev Biol 1994; 165:299-335.
14. Manandhar G, Sutovsky P, Joshi HC et al. Centrosome reduction during mouse spermiogenesis. Dev Biol 1998; 203(2):424-434.
15. Manandhar G, Simerly C, Salisbury JL et al. Centriole and centrin degeneration during mouse spermiogenesis. Cell Motil Cytoskeleton 1999; 43(2):137-144.
16. Woolley DM, Fawcett DW. The degeneration and disappearance of the centrioles during the development of the rat spermatozoon. Anat Rec 1973; 177:289-301.
17. Manandhar G, Schatten G. Centrosome reduction during rhesus monkey spermiogenesis: Tubulin, centrin, and centriole degeneration. Mol Reprod Dev 2000; 56:502-511.
18. Manandhar G, Simerly C, Schatten G. Highly degenerated distal centrioles in rhesus and human spermatozoa. Hum Reprod 2000; 15(2):256-263.
19. Calarco PG. Centrosome precursors in the acentriolar mouse oocyte. Microsc Res Tech 2000; 49(5):428-434.
20. Calarco PG, Siebert M, Hubble R et al. Centrosomal development in early mouse embryos as defined by an autoantibody against periocentriolar material. Cell 1983; 35:621-629.

21. Meng XQ, Fan HY, Zhong ZS et al. Localization of g-tubulin in mouse eggs during meiotic maturation, fertilization, and early embryonic development. J Reprod Dev 2004; 50(1):97-105.
22. Tang CJ, Hu HM, Tang TK. NuMA expression and function in mouse oocytes and early embryos. J Biomed Sci 2004; 11(3):370-376.
23. Lee J, Miyano T, Moor RM. Spindle formation and dynamics of gamma-tubulin and nuclear mitotic apparatus protein distribution during meiosis in pig and mouse oocytes. Biol Reprod 2000; 62(5):1184-1192.
24. Can A, Semiz O, Cinar O. Centrosome and microtubule dynamics during early stages of meiosis in mouse oocytes. Mol Hum Reprod 2003; 9(12):749-756.
25. Messinger SM, Albertini DF. Centrosome and microtubule dynamics during meiotic progression in the mouse oocyte. J Cell Sci 1991; 100(Pt 2):289-298.
26. Carabatsos MJ, Combelles CM, Messinger SM et al. Sorting and reorganization of centrosomes during oocyte maturation in the mouse. Microsc Res Tech 2000; 49(5):435-44.
27. Crozet N, Dahirel M, Chesne P. Centrosome inheritance in sheep zygotes: Centrioles are contributed by the sperm. Microsc Res Tech 2000; 49(5):445-450.
28. Sathananthan AH, Selvaraj K, Trounson A. Fine structure of human oogonia in the foetal ovary. Mol Cell Endocrinol 2000; 161(1-2):3-8.
29. Battaglia DE, Klein NA, Soules MR. Changes in centrosomal domains during meiotic maturation in the human oocyte. Mol Hum Reprod 1996; 2(11):845-851.
30. Sathananthan AH, Ratnasooriya WD, de Silva PK et al. Characterization of human gamete centrosomes for assisted reproduction. Ital J Anat Embryol 2001; 106(2 Suppl 2):61-73.
31. Sutovsky P, Schatten G. Paternal contributions to the mammalian zygote: Fertilization after sperm-egg fusion. Int Rev Cytol 2000; 195:1-65.
32. Sathananthan AH, Ratnam SS, Ng SC et al. The sperm centriole: Its inheritance, replication and perpetuation in early human embryos. Hum Reprod 1996; 11(2):345-356.
33. Palermo GD, Colombero LT, Rosenwaks Z. The human sperm centrosome is responsible for normal syngamy and early embryonic development. Rev Reprod 1997; 2(1):19-27.
34. Nakamura S, Terada Y, Horiuchi T et al. Analysis of the human sperm centrosomal function and the oocyte activation ability in a case of globozoospermia, by ICSI into bovine oocytes. Hum Reprod 2002; 17(11):2930-2934.
35. Nakamura S, Terada Y, Rawe VY et al. A trial to restore defective human sperm centrosomal function. Hum Reprod 2005; 20(7):1933-1937.
36. Simerly C, Zoran SS, Payne C et al. Biparental inheritance of gamma-tubulin during human fertilization: Molecular reconstitution of functional zygotic centrosomes in inseminated human oocytes and in cell-free extracts nucleated by human sperm. Mol Biol Cell 1999; 10(9):2955-2969.
37. Hewitson L, Simerly CR, Schatten G. Fate of sperm components during assisted reproduction: Implications for infertility. Hum Fertil (Camb) 2002; 5(3):110-116.
38. Crozet N. Behavior of the sperm centriole during sheep oocyte fertilization. Eur J Cell Biol 1990; 53(2):326-332.
39. Szollosi D, Ozil JP. De novo formation of centrioles in parthenogenetically activated, diploidized rabbit embryos. Biol Cell 72(1-2):61-66.
40. Shin MR, Kim NH. Maternal gamma (gamma)-tubulin is involved in microtubule reorganization during bovine fertilization and parthenogenesis. Mol Reprod Dev 2003; 64(4):438-445.
41. Sathananthan AH, Tatham B, Dharmawardena V et al. Inheritance of sperm centrioles and centrosomes in bovine embryos. Arch Androl 1997; 38(1):37-48.
42. Kim NH, Simerly C, Funahashi H et al. Microtubule organization in porcine oocytes during fertilization and parthenogenesis. Biol Reprod 1996; 54(6):1397-1404.
43. Tremoleda JL, Van Haeften T, Stout TA et al. Cytoskeleton and chromatin reorganization in horse oocytes following intracytoplasmic sperm injection: Patterns associated with normal and defective fertilization. Biol Reprod 2003; 69(1):186-194.
44. Terada Y, Simerly CR, Hewitson L et al. Sperm aster formation and pronuclear decondensation during rabbit fertilization and development of a functional assay for human sperm. Biol Reprod 2000; 62(3):557-563.
45. Calarco-Gillam PD, Siebert MC, Hubble R et al. Centrosome development in early mouse embryos as defined by an autoantibody against pericentriolar material. Cell 1983; 35(3 Pt 2):621-629.
46. Abumuslimov SS, Nadezhdina ES, Chentsov IS. An electron microscopic study of centriole and centrosome morphogenesis in the early development of the mouse. Tsitologiia 1994; 36(11):1054-1061.
47. Schatten H, Schatten G, Mazia D et al. Behavior of centrosomes during fertilization and cell division in mouse oocytes and in sea urchin eggs. Proc Natl Acad Sci USA 1986; 83(1):105-109.
48. Schatten G, Simerly C, Schatten H. Maternal inheritance of centrosomes in mammals? Studies on parthenogenesis and polyspermy in mice. Proc Natl Acad Sci USA 1991; 88(15):6785-6789.

49. Palacios MJ, Joshi HC, Simerly C et al. g-tubulin reorganization during mouse fertilization and early development. J Cell Sci 1993; 104(Pt 2):383-289.
50. Schatten H, Prather RS, Sun QY. The significance of mitochondria for embryo development in cloned farm animals. Mitochondrion 2005; 5(Issue 5):303-321.
51. Simerly C, Navara CS, Hyun SH et al. Embryogenesis and blastocyst development after somatic cell nuclear transfer in nonhuman primates: Overcoming defects caused by meiotic spindle extraction. Dev Biol 2004; 276(2):237-252.
52. Navara CS, First N, Schatten G et al. Microtubule organization in the cow during fertilization, polyspermy, parthenogenesis, and nuclear transfer: The role of the sperm aster. Dev Biol 1994; 162(1):29-40.
53. Ruddock NT, Machaty Z, Milanick M et al. Mechanism of intracellular pH increase during parthenogenetic activation of in vitro matured porcine oocytes. Biol Reprod 2000a; 63:488-492.
54. Ruddock NT, Machaty Z, Prather RS. Intracellular pH increase accompanies parthenogenetic activation of porcine, bovine and murine oocytes. Reprod Fertil Dev 2000b; 12:201-207.
55. Schatten H, Walter M, Biessmann H et al. Activation of maternal centrosomes in unfertilized sea urchin eggs. Cell Motil Cytoskel 1992; 23:61-70.
56. Saredi A, Howard, Compton DA. Phosphorylation regulates the assembly of NuMA in a mammalian mitotic extract. J Cell Sci 1997; 110:1287-1297.
57. Merdes A, Cleveland DA. The role of NuMA in the interphase nucleus. J Cell Sci 1998; 111:71-9.
58. Gehmlich K, Haren L, Merdes A. Cyclin B degradation leads to NuMA release from dynein/dynactin and from spindle poles. EMBO Rep 2004; 5:97-103.
59. Shin MR, Park SW, Shim H et al. Nuclear and microtubule reorganization in nuclear-transferred bovine embryos. Mol Reprod Dev 2002; 62(1):74-82.
60. Yin XJ, Cho SK, Park MR et al. Nuclear remodelling and the developmental potential of nuclear transferred porcine oocytes under delayed-activated conditions. Zygote 2003; 11(2):167-174.
61. Sun QY, Schatten H. Multiple roles of NuMA in vertebrate cells: Review of an intriguing multifunctional protein. Frontiers in Bioscience 2006; 11:1137-1146.
62. Zhong ZS, Zhang G, Meng XQ et al. Function of donor cell centrosome in intraspecies and interspecies nuclear transfer embryos. Exp Cell Res 2005; 306(1):35-46.
63. Liu ZH, Schatten H, Hao YH et al. The nuclear mitotic apparatus (NuMA) protein is contributed by the donor cell nucleus in cloned porcine embryos. Front Biosci 2006; 11:1945-1957.
64. Simerly C, Dominko T, Navara CS. Molecular correlates of primate nuclear transfer failures. Science 2003; 300(5617):297.
65. Meng L, Ely JJ, Stouffer RL et al. Rhesus monkeys produced by nuclear transfer. Biol Reprod 1997; 454-459.
66. Wolf DP. Assisted reproductive technologies in rhesus macaques. Reprod Biol Endocrinol 2004; 2:37.
67. Schramm RD, Paprocki AM. Strategiesfor the production of genetically identical monkeys by embryo splitting. Reprod Biol Endocrinol 2004; 2:38.
68. Salisbury JL.Centrin, centrosomes, and mitotic spindle poles. Curr Opin Cell Biol 1995; 7:39-45.
69. Manandhar G, Feng D, Yi YJ et al. Centrosomal protein centrin is not detectable during early preimplantation development but reappears during late blastocyst stage in porcine embryos. Reproduction 2006; in press.
70. Hiendleder S. Mitochondrial DNA inheritance after SCNT. In: Sutovsky P, ed. Somatic cell nuclear transfer. Georgetown: Landes Bioscience, New York: Springer Science+Business Media, 2006: 8.
71. Delattre M, Gonczy P. The arithmetic of centrosome biogenesis. J Cell Sci 2004; 117(Pt 9):1619-1930.
72. Higginbotham H, Bielas S, Tanaka T et al. Transgenic mouse line with green-fluorescent protein-labeled Centrin 2 allows visualization of the centrosome in living cells. Transgenic Res 2004; 13(2):155-164.
73. D'Assoro AB, Stivala F, Barrett S. GFP-centrin as a marker for centriole dynamics in the human breast cancer cell line MCF-7. Ital J Anat Embryol 2001; 106(2 Suppl 1):103-110.
74. Danowski BA, Khodjakov A, Wadsworth P. Centrosome behavior in motile HGF-treated PtK2 cells expressing GFP-g tubulin. Cell Motil Cytoskeleton 50(2):59-68.

Chapter 5

Developmental, Behavioral, and Physiological Phenotype of Cloned Mice

Kellie L.K. Tamashiro,* Randall R. Sakai, Yukiko Yamazaki,
Teruhiko Wakayama and Ryuzo Yanagimachi

Abstract

Cloning from adult somatic cells has been successful in at least ten species. Although generating viable cloned mammals from adult cells is technically feasible, prenatal and perinatal mortality is high and live cloned offspring have had health problems. This chapter summarizes the health consequences of cloning in mice and discusses possible mechanisms through which these conditions may arise. These studies have further significance as other assisted reproductive techniques (ART) also involve some of the same procedures used in cloning, and there are some reports that offspring generated by ART display aberrant phenotypes as well. At the moment, the long-term consequences of mammalian cloning remain poorly characterized. Data available thus far suggest that we should use this technology with great caution until numerous questions are addressed and answered.

Introduction

It was once believed that cloning mammals from differentiated adult somatic cells was impossible,[1] but we now know that this is not the case. In 1997, Ian Wilmut and colleagues in Scotland reported that they successfully generated a cloned mammal, a sheep named Dolly, using nuclear transfer of nuclei from an adult mammary gland cell.[2] Dolly was the only sheep generated from over 250 attempts and efforts to replicate that feat failed, casting doubt as to whether Dolly was indeed a clone. A year later, two independent groups, using two different species, confirmed that mammalian cloning using an adult somatic cell was indeed possible.[3,4] Since those first reports, other species have been cloned from adult somatic cells including pig,[5] goat,[6] cat,[7] rabbit,[8] horse,[9] rat,[10] and dog.[11]

Potential Applications

The introduction of nuclear transfer as a viable method of generating genetically identical offspring has opened numerous avenues for application of the technology in basic and clinical sciences. Cloning technology provides an excellent tool to study such processes as nuclear reprogramming, gene activation, and development. Biotechnological applications include the ability to generate transgenic livestock, not only by random gene addition, but also by knock-in or knockout techniques[12] previously restricted to the mouse using embryonic stem (ES) cell techniques. As an example, human factor IX, a clotting factor, can now be produced in sheep milk and isolated for human pharmaceutical use.[13] Livestock can be genetically modified for

*Corresponding Author: Kellie L.K. Tamashiro—Department of Psychiatry and Behavioral
Sciences, Johns Hopkins University School of Medicine, 720 Rutland Avenue, Ross 618,
Baltimore, Maryland 21205, U.S.A. Email: ktamashiro@jhmi.edu

Somatic Cell Nuclear Transfer, edited by Peter Sutovsky. ©2007 Landes Bioscience
and Springer Science+Business Media.

disease-resistance or a higher, better quality yield for human consumption. Replacement tissues for xenotransplantation therapy can be genetically altered to circumvent organ rejection complications.[14,15]

Cloning by transfer of adult somatic cells is also considered the most extreme version of assisted reproduction. It may be possible to use cloning for the propagation of endangered species[16] and beloved family pets such as cats and dogs.[7,11,17,18] The successful application of cloning technology to produce human offspring has been claimed by some;[19] however, there is no evidence that it has indeed occurred or is even possible.[20] Nevertheless, the technology will continue to move forward and the ability to generate a human clone may one day be possible.

Although the efficiency of cloning in all species is extremely low,[21] the cloning technology will improve as scientists determine the optimal conditions for cloning. As with all new technologies, it is important to evaluate the potential short- and long-term consequences prior to its widespread use. In this review, we summarize the studies that have been conducted in mice that begin to document the consequences of mammalian cloning from adult somatic cells. In selecting the mouse as our animal model, we took advantage of its easy handling, well-characterized behavior and genetics, as well as its short lifespan which permits longitudinal studies of these animals in a relatively short time. While the phenotype of cloned offspring of various species has been published, it is important to note that in some cases the results have been mixed, suggesting that the consequences of cloning are highly unpredictable and may depend, in part, on species and nuclear transfer technique used. This further indicates that comprehensive, long-term studies of cloned offspring are necessary.

Additionally, while some altered phenotypes in cloned animals may be attributable to the cloning process itself, additional influences may arise from inter-laboratory differences in technical proficiency. Indeed, some studies report phenotypic variations in animals generated from both normal and reconstructed embryos exposed to in vitro culture conditions.[22-26] Thus, it is critical that the appropriate control groups are included in order to differentiate clone-specific phenotypes from those resulting from technical aspects of the technique itself.[27,28] Of equal, or perhaps greater, significance is the fact that many of the technical manipulations involved in cloning are also used in human assisted reproductive techniques (ART) including in vitro fertilization (IVF), intracytoplasmic sperm injection (ICSI), exchange of cytoplasm between oocytes and in vitro culture. These techniques have been used extensively in humans for a short period of time; Louise Brown, the first IVF baby is currently in her mid-twenties, however the long-term consequences of ART have not been determined. Perturbations in development resulting from in vitro manipulation of embryos have been documented in sheep, cattle and mice[24,29] suggesting that these techniques are not innocuous to offspring. Systematic longitudinal studies must be done, using cloning and ART, and we must be careful in dissecting the effects of these techniques by including control groups to determine the contribution of such manipulations to any aberrations in offspring.

Methods

For the studies discussed here, we generated female B6C3F1 cloned mice by the "Honolulu Technique" and examined the behavioral and physiological consequences of cloning on the resulting offspring.[4] Briefly, female mouse clones were generated by microinjection of cumulus cell nuclei from adult (8-10 week old) B6C3F1 (C57BL/6 x C3H/He) hybrid mice into enucleated oocytes collected from adult (8-10 week old) B6D2F1 (C57BL/6 x DBA/2) mice. Preimplantation embryos were transferred into pseudo-pregnant CD-1 surrogate mothers. Pups were delivered at 19.5 days *post coitum* (d.p.c.) by Caesarean section and placed with the litters of lactating CD-1 foster mothers to be raised.

In addition, we included two control groups. The first group included age- and strain-matched female mice generated by natural mating ("STOCK" controls). The second control group ("IVEM" controls) also consisted of age- and strain-matched female mice. These animals were generated to control for several of the in vitro manipulations and procedures that the cloned embryos were subjected to including in vitro culture until the 2-cell stage and embryo transfer into surrogate mothers. IVEM mice were delivered by Caesarean section at 19.5 d.p.c. and the pups were cross-fostered to lactating CD-1 foster mothers.

Cloning the Mouse

"Cumulina" was the first mouse to be cloned from an adult somatic cell, a cumulus cell.[4] Since this first report, others have followed in producing cloned mice from embryonic stem (ES) cells,[30-33] fetal neurons,[34] fibroblasts,[35-38] and immature Sertoli cells.[37,38] It is important to point out that the cloned mice produced in these studies were generated by piezo-driven microinjection of isolated cell nuclei. This is in contrast to other studies, including those that evaluated the phenotype of cloned cattle, which used whole cell injection and electrofusion to generate cloned embryos. There are inherent differences between the two protocols including the degree of mechanical trauma to the oocyte and reconstructed embryo and the amount of cytoplasmic contamination by contents of the injected somatic cell as well as by a small amount of culture media. The significance of cytoplasmic contamination has not yet been determined.

The method of activation of the resulting embryo differs between the "Honolulu Technique" and the electrofusion method. While the "Honolulu Technique" relies on chemical activation by Sr^{2+} and a delay between nuclear transfer and activation, the electrofusion method involves simultaneous activation by the electric pulse during fusion of the donor cell with the enucleated oocyte. It has been demonstrated that the method of activation does not have a significant effect on postimplantation embryo development or the birth of live offspring;[39] however, whether it produces differences in the phenotype of offspring is not known.

The differences between these two protocols have been addressed[37,38] but have not yet been investigated thoroughly, particularly in regard to the potential long-term consequences on the phenotype of offspring. An additional consideration is the effects of these procedures (i.e., microinjection, artificial activation, in vitro culture) on the development and phenotype of offspring since these are the same procedures commonly used in ART.

Embryonic and Prenatal Development

The success rate (defined as the percentage of reconstructed embryos that develop to term) of mouse cloning is very low and currently remains at about 1% for cumulus cells and 2-3% for embryonic stem (ES) cells. However, differences between the use of these two cell types become apparent when progress to successive developmental stages is examined in detail (Fig. 1).[30,35,36] With cumulus cells, 55% of cloned embryos developed in vitro to the morula/blastocyst stage. Evaluation of in utero development beyond the morula/blastocyst stage was made at the time of Cesarean section, 16 days after embryo transfer; 35% of cloned embryos appeared to have implanted, and 1% developed to live pups, with no dead fetuses.

In contrast, only 30% of cloned embryos completed preimplantation development when ES cells were used as nuclear donors. In utero, 20% showed evidence of implantation, 4% developed into a fetus, and 2% were alive at term. Many placentae (without fetuses) and dead fetuses arrested at 15-17 d.p.c. were observed on Cesarean section. This suggests that the specific restriction points within a successful cloning process may vary with the source of cells. We believe that the mouse is a preferred organism to systematically study the parameters governing the cloning phenomenon.

Placental Abnormalities

Placentomegaly is one striking and consistent characteristic of cloned mice.[35,36,40-43] In cloned mice, placentas are enlarged by approximately two- to three-fold over those of controls, irrespective of gender of clone, nuclear donor source or nuclear transfer protocol.[35-38] Figure 2 shows the placenta weights of female B6C3F1 stock, IVEM and cloned mice (Fig. 2). Abnormal placentas have also been noted in cattle[44] and are thus not unique to cloned mice.

Placental abnormalities in humans have been associated with adverse effects on growth and development of the fetus. In light of this, identifying the pathways and genes involved in normal placentation is important. The mechanisms for larger placental size have been more

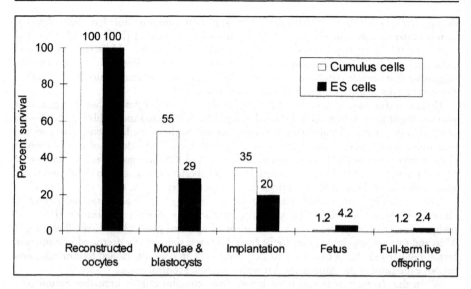

Figure 1. Survival of cloned embryos at various stages of development (adapted from ref. 30). The number of oocytes that survived nuclear transfer of cumulus cells or embryonic stem (ES) cell nuclei is represented as 100%. Advancement to the morula/blastocyst stage occurred over 3.5 days in vitro. At 19.5 d.p.c., cesarean section is performed and the number of implantation sites, stillborn fetuses and live fetuses are noted. The "fetus" stage includes both still-born and live fetuses.

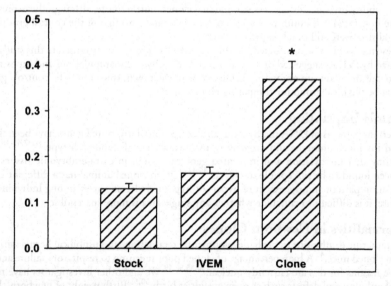

Figure 2. Placental hypertrophy in clones. Typical term placentas in cloned mice are 2- to 3-times heavier than those of control mice. * $P < 0.05$ vs. IVEM and stock.

thoroughly studied in cloned mice and have been attributed to hypertrophy of the basal layer, spongiotrophoblasts, giant trophoblasts and glycogen cells.[41] In addition, placental zonation is disrupted and is characterized by interdigitation of the labyrinthine-basal layer boundary and disorganization of the labyrinthine layer.[41] Similar placental abnormalities have been observed in mice produced by other in vitro micromanipulation techniques [ICSI, round spermatid injection (ROSI), aggregation chimera, and pronuclear exchange].[42] However, only cloned mice exhibited basal layer expansion associated with marked proliferation of glycogen cells suggesting that nuclear transfer is responsible for these specific characteristics.[42] The mechanisms for placental hypertrophy in cloned mice remain unclear.

Recent studies have determined that methylation of *Spalt-like gene 3 (Sall3)*, a gene that plays an important role in nervous system development, is increased and highly correlated with larger placentas in cloned mice. Those studies reported that hypermethylation was not dependent upon cloned mouse gender, strain, or donor cell used.[45] The degree of methylation at *Sall3* in mice generated by in vitro fertilization (IVF) and ICSI was not different from that of naturally mated controls. These data suggest that this aberration is specific to clones and is not species specific, and thus may result from the nuclear transfer process itself.

The expression of some imprinted and nonimprinted genes in placentas and fetuses of cloned mice was compared to control mice produced by in vitro fertilization (IVF). At midgestation (E12.5), placental expression of two imprinted genes (*Meg1/Grb10*) and *Peg1/Mest*) and two nonimprinted genes (*Igfbp2 and Esx1*) was significantly decreased in immature Sertoli cell-derived clones compared to IVF controls.[46] In contrast, cloned fetuses exhibited mRNA expression of the same genes that were within the control range.

When the placentas of cloned mice derived from cumulus cells or immature Sertoli cells were examined at full term, the levels of 3 imprinted genes (*Peg1/Mest, Meg1/Grb10*, and *Meg3/Gtl2*) and 4 nonimprinted genes (*Igfbp2, Igfbp6, Vegfr2/Flk1*, and *Esx1*) were lower than those of IVF controls, regardless of donor cell source. Three genes (*Igf2, H19*, and *Igf2r*) did not exhibit significant difference in clones; however, it was interesting to note that the variability among clones appeared high suggesting that there were differences within the clone groups. Therefore, while some clones may have gene expression similar to that of controls there is still high variability in expression between individual clones. Furthermore, although the expression of some imprinted and nonimprinted genes was comparable to that of the control group, the placental sizes were still much larger for clones.

These results must be considered carefully; the control group that was used in this study was generated by IVF. As suggested previously, some of the physical manipulations and exposure to in vitro culture media are not without side effects of their own; thus, the IVF "control" group may not be the ideal group for comparing cloned mice.

Genomic Imprinting

Aberrant patterns of DNA methylation and expression of imprinted genes have been documented in cloned embryos and progeny in mice regardless of donor cell type.[31,32,45-48] It is interesting that methylation and imprinted gene expression in extra-embryonic tissues also have been found to be susceptible to defects.[36,37,46] Each cloned animal has a different DNA methylation pattern and the extent of hyper- or hypomethylation varies among individuals;[47] therefore, it is difficult to conclude whether any single area is more susceptible.

Abnormalities in Newborn Clones

Intrauterine death prior to birth, developmental retardation, and umbilical hernia are common in cloned mice.[42] A high percentage of cloned pups succumb to respiratory failure at birth and the reasons for this are currently unclear.[30,35,36,49] We and other investigators have noted other developmental defects such as open eyelids at birth.[50,51] Birth weight of newborn clones was not different from that of pups derived by IVF or round spermatid injection (ROSI); however, no STOCK control was included in this study.[42] We have observed increased birth weights of cloned and IVEM mice relative to STOCK control mice,[24] and these observations are consistent with those in cloned and in vitro manipulated cattle and sheep.[52]

Despite the high incidence of pre and perinatal death of cloned mice, we have begun comprehensive studies to examine postnatal development, behavior and phenotype of cloned mice. Although ES cells have been used in many nuclear transfer studies most studies of the long-term consequences of cloning on the phenotype of offspring have been conducted using mice derived from nuclear transfer of somatic cell nuclei. Therefore, this review will focus on cloned mice derived from adult somatic cells.

Pre-Weaning Development of Cloned Mice

The early postnatal development of mice can be assessed by a battery of tests collectively known as the Fox Battery of developmental milestones.[53] The Fox Battery consists of a set of behaviors that each appear at different time points throughout neonatal development. We found no differences in these preweaning developmental assays with the exception of negative geotaxis, ear twitching and eye opening.[54] Although the appearance of these behaviors was slightly delayed in cloned mice compared to controls, the mean day of appearance fell well within the range established for normal mice. Furthermore, the cloned mice performed comparably to control mice in subsequent behavioral tests and did not appear to be adversely affected by the delayed appearance of those milestones. Since animals in these studies were only examined through 6 months of age, longer-term behavioral studies need to be conducted to ascertain that there are no adverse consequences that may emerge over the long-term.

Behavior of Cloned Mice

We used a well-characterized and widely accepted behavioral task, the Morris water maze, to evaluate learning and memory of cloned mice.[55] Both clones and IVEM controls successfully completed the task, finding the submerged platform with a shorter latency over consecutive days of testing. Additionally, there were no differences between the groups, suggesting that both groups were able to use information obtained from previous trials to find the platform faster on subsequent days. Our data also suggests that both groups were employing a spatial learning strategy as indicated by the increased amount of time spent in the quadrant where they had been trained to find the platform. Furthermore, when the position of the submerged platform was changed, the cloned mice navigated to the new platform position and found it with a shorter latency than during the initial acquisition trials. Together these results suggest that cloned mice are capable of completing a spatial learning task and do not have deficits in learning and memory, at least through six months of age.[54]

In order to determine whether cloned mice have normal diurnal activity patterns, 24-hour home cage activity was measured and found to be similar to that of IVEM control mice at all of the time points examined.[54] Some investigators suggest that cloned animals age prematurely.[56,57] In order to provide a rough measure of aging we tested clones' motor skills and abilities. We did not find any deficits in motor coordination, muscle strength, or balance[54] in clones in comparison to their age-matched controls. In sum, cloned mice appear to develop reflexes and other behaviors at the same ages as normal mice and to have normal motor control and coordination.

Taken together, our behavioral data suggests that cloned mice are not significantly different from control mice from birth through six months of age. Longitudinal studies of behavior over the entire lifespan of cloned mice have yet to be conducted to determine whether there are long-term consequences of cloning on offspring behavior. Indeed, some studies have reported indications of premature aging in cloned sheep[56] and shorter lifespan in cloned mice[58] and whether ehavioral deficits occur in concert with these observations remains to be determined. We have examined multiple generations of cloned mice and have not noted any significant deficiencies in any of the developmental or behavioral measures described above.[40]

Aging and Longevity

The use of differentiated adult somatic cells for cloning calls into question the actual age of a cloned animal. Progressive shortening of telomeres during DNA replication and cell division has been associated with cellular aging or senescence.[59] Previous studies have examined telomere lengths as an indirect measure of aging in cloned animals. Sheep were reported to have shorter telomeres suggestive of premature aging[56] while cattle had either age-appropriate[60,61] or longer telomeres.[62] Similarly, cloned pigs have telomere lengths that are comparable to their age-matched controls.[63] The variation of these findings in cloned animals suggests that cloning has different consequences in different species and even within a species.

Cloned mice in our studies do not show behavioral deficits or signs of premature aging.[40,54,64] To further investigate aging in cloned mice, Wakayama et al successfully generated 4 and 6 generations of cloned mice in two independent lines; i.e., clones derived from clones.[40] If cloning mice from adult somatic cells results in short telomeres, serial generations of cloned mice should have telomeres that become progressively shorter. Behaviorally, the six generations of cloned mice did not exhibit signs of premature aging. Telomere lengths were also analyzed in this experiment to provide a molecular measure of aging. Consistent with our behavioral results, no telomere shortening was observed. In fact, telomere size increased with each successive generation.[40] A possible explanation for this occurrence involves the enzyme telomerase which functions to elongate the ends of telomeres thereby stabilizing and protecting the ends of chromosomes during DNA replication and cell division. Although telomerase is not expressed in most somatic cells, telomerase activity was detected in cumulus cells, the donor cells used in these experiments, and suggests that cumulus cells may have longer telomeres to begin with compared to other somatic cells.[40]

In contrast to behavioral and telomere data in cloned mice, Ogonuki et al reported that male B6D2F1 mice cloned from immature Sertoli cells were prone to early death.[58] The high mortality rate in that study was attributed to higher susceptibility to pathological conditions and diseases, mainly pneumonia and hepatic failure, stemming from reduced immunological function. Immuno-suppression has also been reported in cloned cattle[65] and goats[66] suggesting that this condition is not unique to cloned mice. The observations by Ogonuki et al were restricted to male clones of the B6D2F1 mouse strain; 129/Sv and 129 X JF1 Sertoli cell clones did not exhibit the same fate[42] suggesting that early pneumonia-associated death of B6D2F1 Sertoli cell clones was perhaps strain-dependent. Previous studies suggest that telomere shortening may contribute to immunological dysfunctions in mice.[67] Telomere lengths were not measured in the Ogonuki experiment. However, given the degree of variability found in telomere lengths among cloned cattle[60,62,68,69] (which may be dependent upon the donor cell type), it is plausible that the mice in this particular experiment had shortened telomeres leading to premature immune insufficiency.

The variability of results between these studies suggests that longevity in cloned mice warrants further investigation. The female B6C3F1 cloned mice involved in our studies do not die prematurely (Fig. 3).[64] We have similar data in the B6D2F1 strain suggesting that, at least in cloned mice, strain is not a factor (Fig. 3). In fact, the first cloned mouse, Cumulina, lived for over 2.5 years, a very advanced age for a mouse with a 2-year expected lifespan.[70] The longevity of cloned mice may depend on multiple factors including mouse strain, donor cell source and age, as well as the technician's skill during nuclear transfer.[28]

Body Weight and Obesity

Several studies have reported increased body weight in cloned animals, primarily in mice (Fig. 4).[24,42,54] Cloned mice derived from adult cumulus cells have higher body weights and have been found to exhibit characteristics consistent with obesity. Food intake measures showed that cloned mice are not hyperphagic as adults on standard rodent chow.[24] However, we do not know whether clones were hyperphagic prior to the onset of obesity at adulthood. Our preliminary data also suggest that clones may have higher metabolic efficiency (weight gain/g food consumed) since they do not eat significantly more food per gram of body weight than IVEM or STOCK controls, but continue to gain more weight than controls. Energy expenditure has not been measured in cloned mice, and it is possible that energy expenditure is lower in clones prior to weight gain.

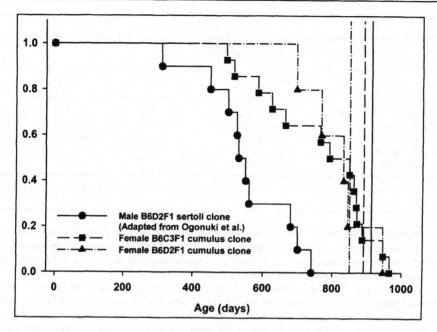

Figure 3. Longevity of female mice cloned from adult cumulus cells.

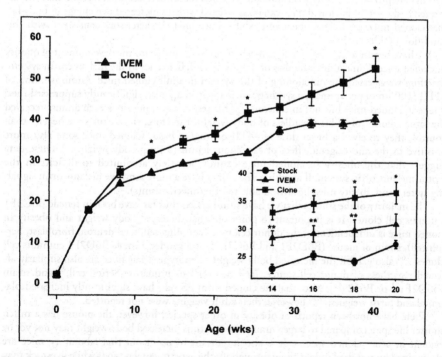

Figure 4. Body weight of cloned mice, IVEM, and F2 offspring. * $P < 0.05$ vs. IVEM and F2; ** $P < 0.05$ vs. F2. Inset, comparison of body weight of clones, IVEM and F2 to age-matched background stock mice at 14-20 weeks of age. * $P < 0.05$ vs. IVEM, F2 and stock; ** $P < 0.05$ vs. stock. Adapted from ref. 24.

Table 1. **Terminal body composition and endocrine parameters of female B5C3F1 mice cloned from adult cumulus cells**

	Stock (n = 7)	IVEM (n = 7)	Clone (n = 9)
Terminal body weight (g)	28.9 ± 0.7	39.3 ± 1.3*	49.7 ± 3.9**
% fat tissue	14.4 ± 1.4	19.5 ± 1.9	25.2 ± 2.8*
% lean tissue	27.5 ± 0.4	34.7 ± 1.9*	32.9 ± 2.1*
Leptin (ng/ml)	2.1 ± 0.5	9.6 ± 1.7*	14.6 ± 1.7**
Insulin (pM)	12.0 ± 2.0	26.0 ± 2.0*	32.0 ± 5.0*

Data are expressed as mean ± S.E.M. * $P < 0.05$ vs. Stock; ** $P < 0.05$ vs. Stock and IVEM. Adapted from ref. 24.

When clones were challenged with an acute hypocaloric challenge, 24-hr food deprivation, they lost the same percentage of body weight as STOCK and IVEM controls. When allowed to refeed, clones responded by increasing their food intake similarly as controls indicating that they defend their body weight in a similar fashion.[24]

Although adult cloned mice were significantly heavier than their age matched controls, this does not imply that they are necessarily obese. Body composition analysis revealed that clones had higher body fat compared to controls, almost twice as much, while the percentage of lean body mass did not differ among the groups.[24] Consistent with this finding, cloned mice were also hyperleptinemic and hyperinsulinemic.[24] These data therefore demonstrate that the increased body weight in cloned mice can be associated with an increased percentage of body fat, that cloned mice are indeed obese and exhibit hormonal characteristics consistent with this condition (Table 1).[24]

We have begun to examine possible mechanisms responsible for the development of obesity in cloned mice. Some animal models of obesity have deficiencies in the melanocortin system, therefore we assessed the functioning of this system in adult cloned mice. Administration of MTII (100 nmol per animal), a synthetic melanocortin agonist, significantly suppressed food intake in cloned mice but not in controls.[24] This result is striking since each animal received the same amount of MTII regardless of body weight; therefore, since clones are heavier than controls they received a lower dose of MTII on mg/kg basis. Cloned mice were also more sensitive to the anorexigenic effect of exogenous leptin (5 μg/g body weight).[24] These data suggest that the obesity in cloned mice cannot be solely attributed to deficits in the leptin-melanocortin system. If anything, the clones have a more sensitive melanocortin signaling system and obesity must be attributable to other mechanism(s).

The majority of the studies that we have summarized thus far have been in female B6C3F1 cumulus cell clones. It is important to point out that increased body weight and obesity in cloned mice is independent of the donor cell type (cumulus cell, fetal neuron, fibroblast, Sertoli cell), strain of mouse (B6D2F1 and B6C3F1), and gender. Female B6D2F1 cumulus cell clones[24,64] also exhibit the increased body weight phenotype. Male mice are also similarly affected regardless of donor cell source, fetal neuron[51] or immature Sertoli cell,[42] and strain [B6D2F1[51] or B6129[42]]. Mice that are cloned from ES cells have significantly increased placental and birth weights;[49,50] however, their adult weights were not reported.

There have not been reports of obesity in other species, however, the mouse has a much shorter life span compared to larger ruminants and thus increased body weight may not yet be evident in other cloned species. It is also important to point out that laboratory mice are maintained on standard rodent chow throughout the experimental periods while livestock may be maintained on diets of varying nutrient composition.

Serial cloning of mice may amplify subtle changes resulting from nuclear transfer. To address this possibility we generated up to 6 generations of cloned mice.[40] Postnatal development and behavioral characteristics in six generations of clones were not different from controls nor were they different between generations of cloned mice.[40] The obese phenotype is maintained, but not enhanced, in successive generations of mice.[40] Interestingly, when a cloned female was mated with a cloned male or a wild type male, the obese and placental hypertrophy phenotypes were not passed on to the resulting offspring.[24] Similarly, mating of mice cloned from ES cells produced offspring that did not display the phenotypic abnormalities that their cloned parents displayed, including placental hypertrophy and open eyelids at birth.[50] Together these data suggest that an epigenetic mechanism may cause aberrant imprinting and/or reprogramming and be responsible for the altered phenotypes,[24,50] and that these alterations are corrected during gametogenesis.

Discussion and Conclusion

We have discussed several studies documenting the phenotype of cloned mice. The overall conclusion that can be drawn from these studies is that cloning is not without side effects that can be serious enough to threaten or compromise the health and survival of cloned offspring. In addition, some aberrations may be evident immediately in newborn clones; however, the results of several studies now suggest that abnormalities may not be manifest until adulthood. In the mouse, this translates into weeks or months, but in the case of domestic species with longer life expectancies, this spans several years. While these studies should ideally be conducted in all cloned species, it is not always cost effective and feasible. The mouse is an excellent model to use in exploring the long-term consequences of cloning.

Systematic longitudinal studies of the phenotype of cloned animals must include appropriate control groups. As more investigators include "manipulated" control groups, it is becoming more evident that in vitro culture and manipulation alone can produce undesirable side effects in offspring. The effects of these procedures must be studied carefully as the results will have consequences on the use of assisted reproductive techniques as well.

Acknowledgements

The authors gratefully thank Drs. Stephen C. Woods and David A. D'Alessio of the University of Cincinnati for their contributions to the obesity and body weight regulation studies and for valuable discussions about the manuscript. This research was supported by: NIH grants DK066596 (R.R.S.) and NS047791 (K.L.K.T.), the Victoria S. and Bradley L. Geist Foundation, the Kosasa Family Foundation (R.Y.), and the University of Cincinnati Mouse Metabolic Phenotyping Center.

References

1. McGrath J, Solter D. Inability of mouse blastomere nuclei transferred to enucleated zygotes to support development in vitro. Science 1984; 226(4680):1317-9.
2. Wilmut I, Schnieke AE, McWhir J et al. Viable offspring derived from fetal and adult mammalian cells. Nature 1997; 385(6619):810-3.
3. Kato Y, Tani T, Sotomaru Y et al. Eight calves cloned from somatic cells of a single adult. Science 1998; 282(5396):2095-8.
4. Wakayama T, Perry AC, Zuccotti M et al. Full-term development of mice from enucleated oocytes injected with cumulus cell nuclei. Nature 1998; 394(6691):369-74.
5. Polejaeva IA, Chen SH, Vaught TD et al. Cloned pigs produced by nuclear transfer from adult somatic cells. Nature 2000; 407(6800):86-90.
6. Keefer CL, Keyston R, Lazaris A et al. Production of cloned goats after nuclear transfer using adult somatic cells. Biol Reprod 2002; 66(1):199-203.
7. Shin T, Kraemer D, Pryor J et al. A cat cloned by nuclear transplantation. Nature 2002; 415(6874):859.
8. Chesne P, Adenot PG, Viglietta C et al. Cloned rabbits produced by nuclear transfer from adult somatic cells. Nat Biotechnol 2002; 20(4):366-9.

9. Galli C, Lagutina I, Crotti G et al. Pregnancy: A cloned horse born to its dam twin. Nature 2003; 424(6949):635.
10. Zhou Q, Renard JP, Le Friec G et al. Generation of fertile cloned rats by regulating oocyte activation. Science 2003; 1088313.
11. Lee BC, Kim MK, Jang G et al. Dogs cloned from adult somatic cells. Nature 2005; 436(7051):641.
12. McCreath KJ, Howcroft J, Campbell KH et al. Production of gene-targeted sheep by nuclear transfer from cultured somatic cells. Nature 2000; 405(6790):1066-9.
13. Schnieke AE, Kind AJ, Ritchie WA et al. Human factor IX transgenic sheep produced by transfer of nuclei from transfected fetal fibroblasts. Science 1997; 278(5346):2130-3.
14. Lai L, Kolber-Simonds D, Park KW et al. Production of alpha-1,3-galactosyltransferase knockout pigs by nuclear transfer cloning. Science 2002; 295(5557):1089-92.
15. Wilmut I, Paterson L. Somatic cell nuclear transfer. Nature 2003; 13(6-10):303-7.
16. Holt WV, Pickard AR, Prather RS. Wildlife conservation and reproductive cloning. Reproduction 2004; 127(3):317-24.
17. Westhusin ME, Burghardt RC, Ruglia JN et al. Potential for cloning dogs. J Reprod Fertil Suppl 2001; 57:287-93.
18. Westhusin M, Hinrichs K, Choi YH et al. Cloning companion animals (horses, cats, and dogs). Cloning Stem Cells 2003; 5(4):301-17.
19. Vogel G. Reproductive biology Cloning: Could humans be next? Science 2001; 291(5505):808b-809.
20. Schatten G, Prather R, Wilmut I. Cloning claim is science fiction, not science. Science 2003; 299(5605):344.
21. Wilmut I, Beaujean N, de Sousa PA et al. Somatic cell nuclear transfer. Nature 2002; 419(6907):583-6.
22. Young LE, Fernandes K, McEvoy TG et al. Epigenetic change in IGF2R is associated with fetal overgrowth after sheep embryo culture. Nat Genet 2001; 27(2):153-4.
23. Boiani M, Eckardt S, Scholer HR et al. Oct4 distribution and level in mouse clones: Consequences for pluripotency. Genes Dev 2002; 16(10):1209-19.
24. Tamashiro KL, Wakayama T, Akutsu H et al. Cloned mice have an obese phenotype not transmitted to their offspring. Nat Med 2002; 8(3):262-7.
25. Ecker DJ, Stein P, Xu Z et al. Long-term effects of culture of preimplantation mouse embryos on behavior. Proc Natl Acad Sci USA 2004; 101(6):1595-600.
26. Fernandez-Gonzalez R, Moreira P, Bilbao A et al. Long-term effect of in vitro culture of mouse embryos with serum on mRNA expression of imprinting genes, development, and behavior. Proc Natl Acad Sci USA 2004; 101(16):5880-5.
27. Solter D. Mammalian cloning: Advances and limitations. Nat Rev Genet 2000; 1(3):199-207.
28. Perry AC, Wakayama T. Untimely ends and new beginnings in mouse cloning. Nat Genet 2002; 30(3):243-4.
29. Young LE, Fairburn HR. Improving the safety of embryo technologies: Possible role of genomic imprinting. Theriogenology 2000; 53(2):627-48.
30. Wakayama T, Rodriguez I, Perry AC et al. Mice cloned from embryonic stem cells. Proc Natl Acad Sci USA 1999; 96(26):14984-9.
31. Humpherys D, Eggan K, Akutsu H et al. Epigenetic instability in ES cells and cloned mice. Science 2001; 293(5527):95-7.
32. Humpherys D, Eggan K, Akutsu H et al. Abnormal gene expression in cloned mice derived from embryonic stem cell and cumulus cell nuclei. Proc Natl Acad Sci USA 2002; 99(20):12889-94.
33. Gao S, McGarry M, Ferrier T et al. Effect of cell confluence on production of cloned mice using an inbred embryonic stem cell line. Biol Reprod 2003; 68(2):595-603.
34. Yamazaki Y, Makino H, Hamaguchi-Hamada K et al. Assessment of the developmental totipotency of neural cells in the cerebral cortex of mouse embryo by nuclear transfer. Proc Natl Acad Sci USA 2001; 98(24):14022-6.
35. Wakayama T, Yanagimachi R. Cloning of male mice from adult tail-tip cells. Nat Genet 1999; 22(2):127-8.
36. Wakayama T, Yanagimachi R. Cloning the laboratory mouse. Semin Cell Dev Biol 1999; 10(3):253-8.
37. Ogura A, Inoue K, Ogonuki N et al. Production of male cloned mice from fresh, cultured, and cryopreserved immature Sertoli cells. Biol Reprod 2000; 62(6):1579-84.
38. Ogura A, Inoue K, Takano K et al. Birth of mice after nuclear transfer by electrofusion using tail tip cells. Mol Reprod Dev 2000; 57(1):55-9.
39. Kishikawa H, Wakayama T, Yanagimachi R. Comparison of oocyte-activating agents for mouse cloning. Cloning Stem Cells 1999; 1(3):153-159.
40. Wakayama T, Shinkai Y, Tamashiro KL et al. Cloning of mice to six generations. Nature 2000; 407(6802):318-9.

41. Tanaka S, Oda M, Toyoshima Y et al. Placentomegaly in cloned mouse concepti caused by expansion of the spongiotrophoblast layer. Biol Reprod 2001; 65(6):1813-21.
42. Ogura A, Inoue K, Ogonuki N et al. Phenotypic effects of somatic cell cloning in the mouse. Cloning Stem Cells 2002; 4(4):397-405.
43. Singh U, Fohn LE, Wakayama T et al. Different molecular mechanisms underlie placental overgrowth phenotypes caused by interspecies hybridization, cloning, and Esx1 mutation. Dev Dyn 2004; 230(1):149-64.
44. Cibelli JB, Stice SL, Golueke PJ et al. Cloned transgenic calves produced from nonquiescent fetal fibroblasts. Science 1998; 280(5367):1256-8.
45. Ohgane J, Wakayama T, Senda S et al. The Sall3 locus is an epigenetic hotspot of aberrant DNA methylation associated with placentomegaly of cloned mice. Genes Cells 2004; 9(3):253-60.
46. Inoue K, Kohda T, Lee J et al. Faithful expression of imprinted genes in cloned mice. Science 2002; 295(5553):297.
47. Ohgane J, Wakayama T, Kogo Y et al. DNA methylation variation in cloned mice. Genesis 2001; 30(2):45-50.
48. Mann MR, Chung YG, Nolen LD et al. Disruption of imprinted gene methylation and expression in cloned preimplantation stage mouse embryos. Biol Reprod 2003; 69(3):902-14.
49. Eggan K, Akutsu H, Loring J et al. Hybrid vigor, fetal overgrowth, and viability of mice derived by nuclear cloning and tetraploid embryo complementation. Proc Natl Acad Sci USA 2001; 98(11):6209-14.
50. Shimozawa N, Ono Y, Kimoto S et al. Abnormalities in cloned mice are not transmitted to the progeny. Genesis 2002; 34(3):203-7.
51. Tamashiro KL, Sakai RR, Yamazaki Y et al. Health consequences of cloning mice. Health Consequences of Cloning: A Inui, Taylor and Francis Books. 2005:1-16.
52. Young LE, Sinclair KD, Wilmut I. Large offspring syndrome in cattle and sheep. Rev Reprod 1998; 3(3):155-63.
53. Fox WM. Reflex-ontogeny and behavioural development of the mouse. Anim Behav 1965; 13(2):234-41.
54. Tamashiro KL, Wakayama T, Blanchard RJ et al. Postnatal growth and behavioral development of mice cloned from adult cumulus cells. Biol Reprod 2000; 63(1):328-34.
55. Morris RGM. Spatial localization does not require the presence of local cues. Learn Motiv 1981; 12:239-260.
56. Shiels PG, Kind AJ, Campbell KH et al. Analysis of telomere lengths in cloned sheep. Nature 1999; 399(6734):316-7.
57. Kuhholzer-Cabot B, Brem G. Aging of animals produced by somatic cell nuclear transfer. Exp Gerontol 2002; 37(12):1317-23.
58. Ogonuki N, Inoue K, Yamamoto Y et al. Early death of mice cloned from somatic cells. Nat Genet 2002; 30(3):253-4.
59. Shay JW, Wright WE. Hayflick, his limit, and cellular ageing. Nat Rev Mol Cell Biol 2000; 1(1):72-6.
60. Tian XC, Xu J, Yang X. Normal telomere lengths found in cloned cattle. Nat Genet 2000; 26(3):272-3.
61. Kubota C, Tian XC, Yang X. Serial bull cloning by somatic cell nuclear transfer. Nat Biotechnol 2004; 22:693-694.
62. Lanza RP, Cibelli JB, Blackwell C et al. Extension of cell life-span and telomere length in animals cloned from senescent somatic cells. Science 2000; 288(5466):665-9.
63. Jiang L, Carter B, Xu J et al. Telomere lengths in cloned transgenic pigs. Biol Reprod 2004.
64. Tamashiro KL, Wakayama T, Yamazaki Y et al. Phenotype of cloned mice: Development, behavior and physiology. Experimental Biology and Medicine 2003; 228(10):1193-1200.
65. Renard JP, Chastant S, Chesne P et al. Lymphoid hypoplasia and somatic cloning. Lancet 1999; 353(9163):1489-91.
66. Keefer CL, Baldassarre H, Keyston R et al. Generation of dwarf goat (Capra hircus) clones following nuclear transfer with transfected and nontransfected fetal fibroblasts and in vitro-matured oocytes. Biol Reprod 2001; 64(3):849-56.
67. Blasco MA. Immunosenescence phenotypes in the telomerase knockout mouse. Springer Semin Immunopathol 2002; 24(1):5-85.
68. Betts D, Bordignon V, Hill J et al. Reprogramming of telomerase activity and rebuilding of telomere length in cloned cattle. Proc Natl Acad Sci USA 2001; 98(3):1077-82.
69. Miyashita N, Shiga K, Yonai M et al. Remarkable differences in telomere lengths among cloned cattle derived from different cell types. Biol Reprod 2002; 66(6):1649-55.
70. Aldhous P. The cloned mouse that roared is silenced. Nature 2000; 405:268.

CHAPTER 6

Nucleolar Remodeling in Nuclear Transfer Embryos

Jozef Laurincik* and Poul Maddox-Hyttel

Abstract

Transcription of the ribosomal RNA (rRNA) genes occurs in the nucleolus and results in ribosome biogenesis. The rRNA gene activation and the associated nucleolus formation may be used as a marker for the activation of the embryonic genome in mammalian embryos and, thus serve to evaluate the developmental potential of embryos originating from varied nuclear transfer protocols. In bovine in vivo developed embryos, functional ribosome-synthesizing nucleoli become structurally distinct toward the end of the 4th post-fertilization cell cycle. In embryonic cell nuclear transfer embryos, fully developed nucleoli are not apparent until the 5th cell cycle, whereas in somatic cell nuclear transfer embryos the functional nucleoli emerge already during the 3rd cell cycle. Intergeneric reconstructed embryos produced by the fusion of bovine differentiated somatic cell to a nonactivated ovine cytoplast fail to develop fully functional nucleoli. In bovine in vivo developed embryos, a range of important nucleolar proteins (e.g., topoisomerase I, upstream binding factor and RNA polymerase I, fibrillarin, nucleophosmin and nucleolin) become localized to the nucleolar anlage over several cell cycles. This relocation is completed toward the end of the 4th cell cycle. A substantial proportion of bovine embryos produced by nuclear transfer of embryonic or somatic cells to bovine ooplasts display aberrations in protein localization in one or more blastomers. This information is indicative of underlying aberrations in genomic reprogramming and may help to explain the abnormalities observed in a proportion of fetuses and offspring derived from nuclear transfer embryos.

Introduction

The technique of cloning by nuclear transfer, wherein a cell is transferred into an enucleated oocyte, has now in several mammalian species proven that the oocyte possesses factors that can entirely de- and redifferentiate (reprogram) the genome. According to the current hypothesis, reprogramming is an epigenetic event by which the highly specialized gene expression pattern of the donor cell is more or less completely erased and replaced by an embryo specific expression program. By examining the nuclear architecture and dynamic chromatin characteristics of a cell exposed to these factors in the oocyte cytoplasm, it is possible to gain insight into the underlying molecular changes, and potentially those that are required to, in the future, induce a cell to transdifferentiate. Recent research has uncovered some of the mechanisms involved in the genomic reprogramming exerted by the oocyte during nuclear transfer. Hence, the process involves a series of molecular events controlling gene expression, and has a profound effect on

*Corresponding Author: Jozef Laurincik—Constantine the Philosopher University, Faculty of Natural Sciences, Trieda A. Hlinku, SK-949 74 Nitra, Slovak Republic. Email: Jlaurincik@ukf.sk

Somatic Cell Nuclear Transfer, edited by Peter Sutovsky. ©2007 Landes Bioscience and Springer Science+Business Media.

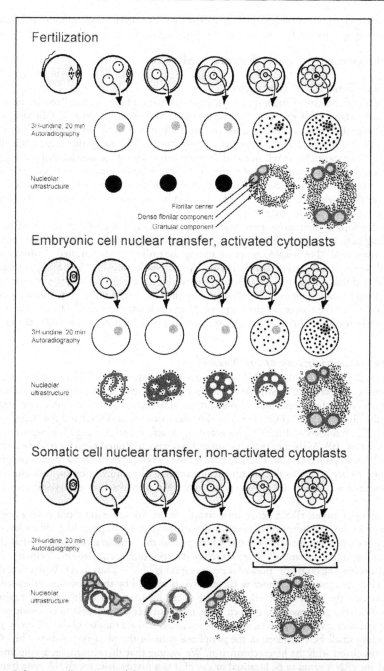

Figure1. Schematic illustration of the nucleolar and extra-nucleolar transcriptional activity as evaluated by 20 min ^{3}H-uridine incubation followed by autoradiography on semi-thin Epon sections; and the corresponding nucleolar ultrastructure during the development of bovine zygotes and embryos after fertilization, embryonic cell nuclear transfer and somatic cell nuclear transfer is shown. See text for details.

nuclear architecture. It is evident that gene expression is silenced during the initial phases of reprogramming, which can be monitored by disassembly of the nucleolus.[1]

Nucleologenesis in Early Bovine Embryos

Maternal-Embryonic Transition

Initial development of mammalian preimplantation embryos is regulated by gene transcripts and polypeptides produced by and stored in the oocyte. However, following one to three cleavage divisions, control of development is taken over by the expression of portions of the embryonic genome, and the maternally derived transcripts and proteins are gradually degraded.[2-4] This transition from maternal to embryonic control of development is a gradual phenomenon. Overall, maternal-embryonic transition in cattle appears to include at least two phases. The first phase includes a low rate of transcription during the first two cell cycles and an even lower rate during the third cycle.[5] The second phase consists of an emphatic transcription initiated during the fourth cycle.[6]

The developing preimplantation embryo has a profound need for synthesis of proteins supporting both housekeeping and cell differentiation. With this in mind, it is not surprising that the ribosomal RNA (rRNA) genes are among the earliest genes activated. A proper activation of these genes is crucial for continued embryonic development.[7,8] Accordingly, it has been demonstrated that bovine zygotes and embryos exhibit a quantitative decrease in protein synthesis from fertilization up to the 8-cell stage, i.e., the time of the major transcriptional activation, upon which the synthesis again increases up to the blastocyst stage.[9] The decrease may be caused by decreasing contents of mRNA or ribosomes in the embryo, or a combination of the two.

Structural Composition of the Nucleolus

The nucleolus is the site of rRNA transcription and formation of the ribosomal subunits. Furthermore, the nucleolus is structurally the most prominent nuclear organelle. At the ultrastructural level, the functional nucleolus is composed of three well-defined sub-compartments, namely the fibrillar centers (FCs), the dense fibrillar component (DFC) and the granular component (GC). A nucleolus displaying these features is referred to as being fibrillo-granular (For a review, see ref. 10). As the preimplantation embryo has a profound need for protein synthesis, a large ribosome pool is crucial. The activation of the rRNA genes follows a specific program in the bovine embryo. Due to the prominent ultrastructural manifestation of nucleolus formation associated with this process, these genes may serve as markers for embryonic gene activation.[11]

In bovine embryos, rRNA genes are activated at the 4th embryonic cell cycle where they participate in the establishment of functional nucleoli (Fig. 1). In preceding cell cycles, different nuclear entities are observed, none of which possess the structural components that characterize the functional nucleolus.[12] They are referred to as "compact nucleolus", "primary nucleolus", "nucleolus-like body" or "nucleolus precursor body".[8,13,14] Thus, it is believed that during initial cell cycles, protein synthesis in the embryo is secured by the ribosomes inherited from the oocyte. At the onset of the 4th cell cycle, nucleolus precursor bodies emerge again after mitosis. With the progression of this cell cycle, these entities develop first an eccentric primary vacuole and later several peripheral secondary vacuoles. Toward the end of the cell cycle, a DFC surrounding small FCs develops in the peripheral zone of the precursor bodies. The FCs are closely associated with the heterochromatin. We assume that this chromatin arrangement allows for the rRNA genes to be localized in the FCs as a prerequisite for rRNA gene transcription. Additionally, patches of DFC form at the inner lining of the large primary vacuoles.[8,12] Subsequently, the DFC and the FCs develop and a presumptive GC consisting of preribosomal particles emerges, occupying the remaining portion of the nucleolus precursor body. Through this process, fibrillo-granular nucleoli emerge. During the 5th cell cycle of bovine embryos, fibrillo-granular nucleoli are formed already at the onset of the cycle indicating that the premeiotic inactivation of the rRNA genes was finally reversed.

Molecular Composition of the Nucleolus

The assignment of the particular steps of ribogenesis to the different nucleolar compartments has been the subject of intense studies. The rDNA and some of the proteins engaged directly in the transcription thereof (e.g., RNA polymerase I, upstream binding factor UBF, promotor selectivity factor SL-1 and topoisomerase I) have been localized to the FC, to the DFC or even to both compartments.[15,16] In brief, the main functions of these proteins are as follows: Topoisomerase I is required for uncoiling of the rDNA to allow for transcription. Activation of rDNA transcription requires the formation of the so-called transcription initiation complex, consisting of RNA polymerase I, several RNA polymerase I-associated factors and at least two transcription factors, the species-specific SL-1 and UBF. This complex binds to rDNA and initiates transcription. Transcription of rDNA is thought to occur at the interface between the FC and the DFC.[17] Consequently, the nascent rRNA is located mainly in the FC and the inner portion of the DFC, together with proteins engaged in the early processing of the rRNA.[18,19] Proteins involved in the later steps of rRNA processing and ribosome subunit formation (e.g., nucleolin and nucleophosmin) are, in turn, located in the DFC and the GC, respectively.[20]

In order to describe the intranuclear localization of the key nucleolar proteins in early bovine embryos,[12] they were labeled with antibodies against topoisomerase I, RNA polymerase I, UBF, fibrillarin, nucleolin and nucleophosmin. It was observed that in bovine zygotes, topoisomerase I, fibrillarin, nucleolin and, in particular, RNA polymerase I of maternal origin were localized to small spherical bodies in the pronuclei in a proportion of the specimens. Antibodies against UBF and nucleophosmin did not label nuclear entities in any zygotes. The same pattern applied to the 2-cell embryos except for the lack of nucleolin labeling. In the 4-cell embryos no nuclear entities were labeled. In the 8-cell embryos, a progressively increasing labeling of all the above mentioned proteins was noticed with the progression of the cell cycle. Initially, labeling was observed in a proportion of the embryonic nuclei in solitary or scattered foci, but late during the cell cycle all embryos exhibited labeling, in which the different proteins were localized to specific compartments. Labeling for topoisomerase I, UBF and RNA polymerase I was found in clusters of foci corresponding with the localization to FCs in the putative nucleoli. Labeling for fibrillarin was found in similar clusters of foci, but also to a lesser extent in the portions between the foci corresponding with the localization to FCs and DFC. Labeling for nucleophosmin and nucleolin was found in shell-like patterns resulting in ring-like confocal sections corresponding with the periphery of the putative nucleoli, presumably in the GC. Similar labeling patterns were noticed in all tentative16-cell embryos.

Nucleologenesis after Nuclear Transfer of Embryonic Cells

Nuclear transfer of cells derived from preimplantation bovine embryos (embryonic cell nuclear transfer embryos; ECNT embryos) has been pursued for production of multiple embryos with identical genomes. In order to elucidate the reprogramming of the rRNA genes in the nuclei of embryonic cells,[21] in vitro produced donor embryos were cultured for 5-6 days. Cytoplasts were produced by bovine oocyte maturation in vitro for 18-20 h followed by cumulus denudation and enucleation within 2 h. At 24 h, the cytoplasts were activated by 5 min incubation in 7% ethanol followed by 5 h culture in 10 μg/mL cycloximide and 5 μg/mL cytochalasin B. Subsequently, individual embryonic cells were transferred into the perivitelline spaces of the activated bovine cytoplasts and the karyoplast-cytoplast complexes were exposed to a double electric pulse of 2.1 kV/cm for 10 μsec to initiate fusion. The reconstructed nuclear transfer embryos were cultured and fixed for further observations at the 1-cell stage (1st cell cycle), the 2-cell stage (2nd cell cycle), the 4-cell stage (3rd cell cycle), the tentative 8-cell stage (4th cell cycle) and the tentative 16-cell stage (5th cell cycle).

Activation of extra-nucleolar and nucleolar transcription, as demonstrated by a marked incorporation of ^{3}H-uridine into RNA following short term (20 min) incubation, was observed during the 4th and 5th cell cycle, respectively (Fig. 1).

The structural composition of the nuclei (Fig. 1) of the one-cell stage nuclear transfer embryos displayed several whorls consisting of fibrillar material surrounded by large electron-dense granules. The nuclei of the 2- and 4-cell embryos displayed structural entities of two basic types: Large clusters of small electron-dense granules and fibrillar bodies displaying multiple vacuoles. Both entities were associated with condensed chromatin. The fibrillar bodies were surrounded by large electron-dense granules. A similar pattern was observed in the nuclei of the tentative 8-cell embryos. However, in these embryos large electron-dense granules covered both the outside of the fibrillar bodies and the inside of their vacuoles. During the 5th cell cycle, some embryos displayed initial stages of formation of fibrillo-granular nucleoli from the spherical fibrillar anlage. On some locations in the nucleoli, small FCs surrounded by a DFC, were observed. Condensed chromatin was attached to the nucleoli at these locations. Numerous electron-dense granules were embedded in other locations within the nucleoli, signaling the formation of GC. In other embryos, however, well developed reticulated fibrillo-granular nucleoli had already developed.

Regarding the molecular composition of the nucleoli, the labeling of fibrillarin, UBF, nucleolin, and nucleophosmin in all 1-cell embryos was localized to large spherical bodies displaying the most intense labeling towards the periphery. RNA polymerase I was localized to small clusters of foci. None of the embryos displayed labeling for topoisomerase I. In all blastomeres of the 2-cell embryos, fibrillarin, UBF, and RNA polymerase I were localized to clusters of foci, whereas nucleolin and nucleophosmin were localized to spherical bodies displaying the most intense labeling towards the periphery. In 4-cell embryos, fibrillarin, nucleolin, and nucleophosmin were localized to the spherical bodies displaying the most intense labeling towards the periphery in all blastomeres in all embryos. Topoisomerase I was localized to clusters of foci or single foci in all cells in all embryos. The same localization was observed for UBF and RNA polymerase I. However, the latter two proteins were only found in a proportion of cells in about half of the embryos. In 8-cell embryos, nucleolin and nucleophosmin were localized to large spherical structures that were most intensely labeled towards the periphery in all cells in all embryos. The remaining proteins could only be localized to a proportion of cells in all embryos, and some embryos completely lacked labeling. In labeled cells, fibrillarin was localized to clusters of intensely labeled foci, whereas UBF, topoisomerase I and RNA polymerase I were localized to small discrete foci that on some occasions formed small clusters. In the 16-cell embryos, nucleolin and nucleophosmin were localized to large spherical structures that were most intensely labeled towards the periphery in all cells in all embryos. The remaining proteins could only be localized to a proportion of cells in all embryos. In labeled cells, fibrillarin was localized to clusters of intensely labeled foci, whereas UBF, topoisomerase I and RNA polymerase I were localized to clusters of small discrete foci.

Nucleologenesis after Somatic Cells Nuclear Transfer

Within the past nine years, viable offspring have been produced by somatic cell nuclear transfer (SCNT) in a number of species. This technique is, however, still hindered by an exceedingly high rate of embryonic, fetal and neonatal mortality. In order to understand the reprogramming of the rRNA genes in nuclei of somatic cells exposed to the ooplasmic factors during the somatic nuclear transfer,[22] bovine oocytes were enucleated with removal of minimal cytoplasmic volume at 18-20 h after the onset of maturation. Individual serum-starved bovine granulosa cells were placed into the perivitelline space of enucleated oocytes at 20-22 h after maturation and the karyoplast-cytoplast complexes were exposed to a double electric pulse of 2.1 kV/cm for 10 μsec in order to initiate their fusion. At 24 h after the onset of maturation (2 h post fusion), the fused karyoplast-cytoplast complexes, i.e., the reconstructed embryos, were activated by a 5 min incubation in 7% ethanol followed by 5 h culture in 10 μg/ml cycloheximide and 5 μg/ml cytochalasin B. The reconstructed embryos were harvested for further observations at the 1-cell stage (1st cell cycle), the 2-cell stage (2nd cell cycle), the 4-cell stage (3rd cell cycle), the tentative 8-cell stage (4th cell cycle) and the tentative 16-cell stage (5th cell cycle).

As demonstrated by a marked incorporation of ^{3}H-uridine into RNA following short term (20 min) incubation, extra-nucleolar as well as nucleolar transcription was initiated already during the 3rd cell cycle (Fig. 1).

In the 1-cell embryos, the most prominent nuclear entities were nucleoli consisting of a meshwork of presumptive DFC with numerous, embedded, presumptive FCs. The nucleoli developed into large shell-like bodies. During the second cell cycle in some embryos, the nuclei presented shell-like nucleolar bodies, sometimes with adhering clusters of electron-dense granules. In other embryos, nuclei contained classical NPBs typical of early bovine embryos, a phenomenon that may indicate complete, i.e., classical NPBs, vs. incomplete, i.e., shell-like nucleolar bodies, genomic reprogramming. In the 4-cell embryos, the nuclei of some blastomeres exhibited large NPBs with one large eccentrically located vacuole and several smaller peripheral vacuoles. Chromatin penetrated the periphery of these bodies. Other blastomeres presented fibrillo-granular nucleoli with FCs, DFC and a well-developed GC. In the subsequent cell cycles, the formation of fully active nucleoli was observed. However, some nuclei exhibited nucleoli with an abnormal connection to condensed chromatin.

With respect to the molecular composition of the nucleoli, topoisomerase I could not be observed until the 5th cell cycle in spite of fibrillo-granular nucleoli being visible already during the 3rd cell cycle. This phenomenon may be due to limited immunocytochemical detection. RNA polymerase I persisted during the observed period of embryonic development; in particular, discrete clusters of foci were observed during the 5th cell cycle, consistent with the existence of prominent fibrillo-granular nucleoli. UBF was observed during the 1st cell cycle and was localized to clusters of foci from the 16-cell stage (5th cell cycle). Many embryos failed to show labeling of UBF even at late developmental stages. Fibrillarin persisted during the observed period of embryonic development; in particular, during the 4th and 5th cell cycle it assumed a pattern consistent with the existence of prominent fibrillo-granular nucleoli. Nucleophosmin was observed for the first time from the 3rd cell cycle. The same applied for nucleolin, except that this protein was also localized to nuclear entities during the 1st cell cycle. During the 4th and 5th cell cycles the labeling patterns for both proteins were consistent with the formation of fibrillo-granular nucleoli.

Intergeneric Somatic Cell Nuclear Transfer Embryos

Since the birth of the first animal created by somatic cell nuclear transfer, the mechanism of genomic reprogramming of the donor nucleus by the oocyte cytoplasm has been an enigma. SCNT offers a unique opportunity to manipulate the nucleo-cytoplasmic interaction allowing analysis of the communication between the two compartments. The reprogramming capacity of the ooplasm, to some degree, appears to be conserved between mammalian species. Hence, intergeneric SCNT has been reported to sustain blastocyst development using bovine ooplasm and nuclei from a range of different animals, including those from porcine and ovine cells.[23,24] By contrast, intergeneric SCNT embryos, created by transferring a porcine or bovine somatic nucleus into ovine ooplasm, did not develop beyond the 16-cell stage.[25]

In order to examine the degree of ooplasm-nucleus compatibility in ruminant intergeneric SCNT embryos,[26] passage 1 bovine granulosa cells were serum starved for 2 days. Ovine cumulus-oocyte complexes were matured in vitro for 18-23 h, cleaned of all cumulus cells and incubated for 5 min in 7.5 mg/ml cytochalasin B. Ova were subsequently enucleated in medium containing cytochalasin B. Then, the zona pellucida was removed enzymatically with 0.5% pronase. Each cytoplast was placed for 1-2 sec in a drop of protein-free SOF-HEPES containing 200 mg/ml of phytohemagglutinin and then moved to a handling medium drop where they were rolled individually onto a single granulosa cell, leading to adhesion of the two membranes. Groups of 10-15 couplets were transferred to a fusion chamber. Each couplet was individually pulsed with a 400 kHz alignment pulse and then two 80 μsec 1.25 kV/cm pulses. Within 30 min of fusion, reconstructed embryos were activated with calcium ionophore, and incubated in 6-dimethylaminopurine (6-DMAP) for 2 hr. The reconstructed zona-free embryos were cultured in vitro and fixed at the early and late 8-cell stages.

The ultrastructure of the bovine-ovine SCNT embryos revealed NPBs with numerous vacuoles characteristic of a typical ruminant NPB during initial nucleolar formation. However, development of fibrillo-granular nucleoli was not observed in any of the embryos. On the other hand, complex structures consisting of long strands of densely packed fibrillar material coated by granules were observed in some late eight-cell embryos indicating tentative achievement of a certain level of nucleolar development followed by disintegration.

Comparative Aspects of Nucleologenesis after SCNT

Several previous studies have demonstrated that the reconstruction of embryos by nuclear transfer has a profound impact on genomic function, which is clearly signaled by the fact that active transcription ceases promptly in one-cell reconstructed bovine embryo.[21,27-29] Reinitiation of major transcriptional activity in the reconstructed embryos may follow different patterns.[30] The major genomic activation of transcription has been reported to occur in bovine embryos reconstructed from embryonic cells and activated cytoplasts during the 3rd[29] or 4th cell cycle.[21] However, in bovine embryos reconstructed from granulose cells and nonactivated cytoplasts, the extra-nucleolar as well as nucleolar transcription may be initiated during the 3rd cell cycle.[22]

The transfer of a donor cell nucleus into an enucleated cytoplast leads to major nucleolar remodeling. The impact of this remodeling on the nucleolus varies. The reconstruction of bovine embryos from embryonic cells and activated cytoplasts may result in a reversal of the nucleolar structure to typical nonvacuolized nucleolus precursor bodies during the first two cell cycles[28] or in remodeling into more irregular and vacuolized fibrillar bodies, which may subsequently turn into vacuolized nucleolus precursor bodies during the 3rd or 4th cycle.[21,31] The latter type of incomplete remodeling was also observed by,[28] using nonactivated cytoplasts. However, the formation of fibrillo-granular nucleoli in ECNT[21] and SCNT[22] embryos observed during the 5th or 3rd cell cycle, respectively, suggests that even if substantial portion of the embryos revealed structural abnormalities, development of functional fibrillo-granular nucleoli in some embryos may still contribute to the production of preribosomal subunits, supporting early embryonic development in nuclear transfer embryos. Intergeneric nuclear transfer embryos originating from bovine granulosa cells fused to nonactivated ovine cytoplasts exhibited both ruminant type NPBs as well as complex structures consisting of long strands of densely packed fibrillar material coated by granules. However, the fact that the bovine fibrillo-granular nucleoli did not form in the bovine-ovine SCNT embryos suggests that despite the similarities in nucleologenesis in both species, interspecific molecular differences exist, or demands for crucial proteins differ between the nuclear and cytoplasmic entities.

The nucleolar protein compartment in bovine embryos produced in vivo[12] is assembled over several cell cycles which culminate in the development of the fully functional nucleolus during the period of rDNA transcription reactivation. Therefore, in 8-cell embryos developed in vivo, the labeling of key nucleolar proteins was localized to the structures corresponding to developed FCs, DFC and GC. In bovine ECNT[21] and SCNT[22] embryos, the investigated proteins were localized to the similar nucleolar structures at the onset of rDNA genome reactivation. However, lack of the labeling of topoisomerase I, UBF, and fibrillarin in one or more blastomeres in a substantial portion of ECNT and SCNT embryos indicate a decreased developmental potential of these embryos with respect to rDNA transcription and production of preribosomal subunits.

Perspectives and Conclusions

Transcriptional activity of the rDNA in mammalian embryos can be regulated at different levels in the embryos reconstructed by nuclear transfer. Some of the factors involved in transcriptional silencing of ribosomal genes during final stages of oocyte growth could also be involved in the remodeling of the donor cell nucleolus after the nuclear transfer. The activity of UBF may be hindered by pocket proteins, expressed in cells entering quiescence upon culture to confluency. The pocket protein p130, in particular, may be involved in the repression of rDNA transcription in bovine and porcine oocytes.[32,33] According to our results observed in oocytes approaching the end of the growth phase, p130 is targeted to the nucleolus, where it becomes colocalized with and may inactivate UBF and thus suppress rRNA synthesis. Concomitantly, transcription of mRNA encoding the RNA polymerase I-associated factor PAF53 (PAF53) is downregulated. At the ultrastructural level, these molecular changes are paralleled by marginalization of the FCs of the oocyte nucleolus. However, whether the same mechanisms apply for rRNA genome silencing in nuclei exposed to the factors of the ooplasm during the SCNT is yet to be investigated.

As mentioned above, mammalian preimplantation embryogenesis is initially dependent upon maternally inherited molecules during the early embryonic cell cycles. As development proceeds and maternally inherited molecules are selectively degraded, the process of embryogenesis becomes dependent upon the expression of genetic information derived from the embryonic genome. As observed in bovine embryos developed in vivo,[12] proteins involved in rDNA gene transcription and rRNA processing were localized into intranuclear entities during the 4th cell cycle. However, the precise intranuclear localization and interplay of proteins of RNA polymerase I transcription and processing machineries during the rRNA genome reactivation has not yet been studied in bovine nuclear transfer embryos. Moreover, it is not known whether in nuclear transfer embryos the proteins of the polymerase I transcription and processing machineries are transcribed and translated de novo or recruited from a maternal protein pool.

In bovine SCNT embryos fibrillo-granular nucleoli were reported to develop as early as the 3rd cell cycle, which is one cell cycle earlier than expected.[22] Moreover, the labeling of proteins of rDNA gene transcription machinery was lacking in a large portion of the embryos during the 3rd and 4th cell cycle. Therefore we can speculate whether this structure is active with respect to rDNA gene transcription. It is possible that pre-rRNAs originating from the oocyte, rather than embryonic pre-rRNAs, can contribute to the development of fibrillo-granular compartment of the embryonic nucleoli before rDNA gene transcription reactivation. However, such hypothesis has never been proven in bovine reconstructed embryos.

The mapping of embryonic rRNA gene activation and nucleolus formation may serve as a marker for the activation of the embryonic genome. Data obtained in our laboratories suggest that differences exist in nucleolus formation between bovine in vivo developed and nuclear transfer embryos. They may indicate other underlying gene expression deviations that lead to abnormalities. It is our hope that the increasing basic cell biological understanding of the deviations in embryonic development resulting form embryo technological procedures will be of value in the continued development of these technologies.

Acknowledgements

This work was supported by the Danish Agricultural and Veterinary Research Council, DFG, NATO, European Commission and the Alexander von Humboldt-foundation.

References

1. Misteli M. A nucleolar disappearing act in somatic cloning. Nature Cell Biol 2003; 5:183-184.
2 Telford NA, Watson AJ, Schultz GA. Transition from maternal to embryonic control in early mammalian development: A comparison of several species. Mol Reprod Dev 1990; 26:90-100.
3 de Sousa PA, Watson AJ, Schultz GA et al. Oogenetic and zygotic gene expression directing early bovine embryogenesis: A review. Mol Reprod Dev 1998; 51:112-121.
4. Watson AJ, Westhusin ME, de Sousa PA et al. Gene expression regulating blastocyst formation. Theriogenology 1999; 51:117-133.
5. Hay-Schmidt A, Viuff D, Hyttel P. Transcription in in vivo produced bovine zygotes and embryos. Theriogenology 1997; 47:215, (Abstract).
6. Camous S, Kopecny V, Fléchon JE. Autoradiographic detection of the earliest stage of [3H]-uridine incorporation into the cow embryo. Biol Cell 1986; 58:195-200.

7. King WA, Niar A, Chartrain I et al. Nucleolus organizer regions and nucleoli in preattachment bovine embryos. J Reprod Fert 1988; 82:87-95.
8. Kopecny V, Fléchon JE, Camous S et al. Nucleologenesis and the onset of transcription in the 8-cell bovine embryo: Fine-structural autoradiographic study. Mol Reprod Dev 1989; 1:79-90.
9. Frei RE, Schulz GA, Chursch RB. Qualitative and quantitative changes in protein synthesis occur at the 8-16-cell stage of embryogenesis in the cow. J Reprod Fert 1989; 86:637-641.
10. Wachtler F, Stahl A. The nucleolus: A structural and functional interpretation. Micron 1993; 24:473-505.
11. Kopecny V, Niemann H. Formation of nuclear micro-architecture in the preimplantation bovine embryo at the onset of transcription: Implication for biotechnology. Theriogenology 1993; 39:109-119.
12. Laurincik J, Schmoll F, Mahabir E et al. Nucleolar proteins and ultrastructure in bovine in vivo developed, in vitro produced and parthenogenetic cleavage-stage embryos. Mol Reprod Dev 2003; 65:73-85.
13. Fakan S, Odartchenko N. Ultrastructural organization of the cell nucleus in early mouse embryo. Biol Cell 1980; 37:211-218.
14. Geuskens M, Alexandre H. Ultrastructural and autoradiographic studies of nucleolar development and rRNA transcription in preimplantation mouse embryos. Cell Differ 1984; 14:125-134.
15. Biggiogera M, Malatesta M, Abolhassani-Dadras S et al. Revealing the unseen: The organizer region of the nucleolus. J Cell Sci 2001; 114:3199-3205.
16. Koberna K, Malinsky J, Pliss A et al. Ribosomal genes in focus: New transcripts label the dense fibrillar components and form clusters indicative od "Christmas trees" in situ. J Cell Biol 2002; 157:743-748.
17. Hozak P, Cook PR, Schofer C et al. Site of transcription of ribosomal RNA and intranucleolar structure in HeLa cells. J Cell Sci 1994; 107:639-648.
18. Ochs RL, Lischwe MA, Spohn WH et al. Fibrillarin: A new protein of the nucleolus identified by autoimmune sera. Biol Cel 1985; 54:123-133.
19. Raska I, Reimer G, Jarnik M et al. Does the synthesis of ribosomal RNA take place within nucleolar fibrillar centers or dense fibrillar component? Biol Cell 1989; 65:79-82.
20. Biggiogera M, Bürki K, Kaufmann SH et al. Nucleolar distribution of proteins B23 and nucleolin in mouse preimplantation embryos as visualized by immunoelectron microscopy. Development 1990; 110:1263-1270.
21. Hyttel P, Laurincik J, Zakhartchenko V et al. Nucleolar protein allocation and ultrastructure in bovine embryos produced by nuclear transfer from embryonic cells. Cloning 2001; 3:69-82.
22. Laurincik J, Zakhartchenko V, Stojkovich M et al. Nucleolar protein allocation and ultrastructure in bovine embryos produced by nuclear transfer from granulosa cells. Mol Reprod Dev 2002; 61:477-487.
23. Dominko T, Mitalipova M, Haley B et al. Bovine oocyte cytoplasm supports development of embryos produced by nuclear transfer of somatic cell nuclei from various species. Biol Reprod 1999; 60:1496-1502.
24. Lee B, Wirtu GG, Damiani P et al. Blastocyst development after intergeneric nuclear transfer of mountain bongo antelope somatic cells into bovine oocytes. Cloning Stem Cells 2003; 5:25-33.
25. Hamilton H, Kleeman D, Rudiger S et al. In vitro development of cross-species nuclear transfer embryos constructed with ovine, bovine, and porcine donor cells and ovine cytoplasts. Theriogenology 2003; 59:255, (Abstract).
26. Hamilton HH, Peura TT, Laurincik J et al. Ovine ooplasm directs initial nucleolar assembly in embryos cloned from ovine, bovine and porcine cells. Mol Reprod Dev 2004; 69:117-125.
27. Kanka J, Fulka Jr J, Fulka J et al. Nuclear transplantation in bovine embryos: Fine structural and autoradiographic studies. Mol Reprod Dev 1991; 29:110-116.
28. Kanka J, Smith SD, Soloy E et al. Nucleolar ultrastructure in bovine nuclear transfer embryos. Mol Reprod Dev 1999; 52:253-263.
29. Smith SD, Soloy E, Kanka J et al. Influence of recipient cytoplasm cell stage on transcription in bovine nucleus transfer embryos. Mol Reprod Dev 1996; 45:444-450.
30. Wrenzycki Ch, Herrmann D, Lucas-Hahn A et al. Messeger RNA expression patterns in bovine embryos derived from in vitro procedures and their implicarions for development. Reprod Fert Dev 2005; 17:23-35.
31. Lavoir MC, Kelk D, Rumph N et al. Transcription and translation in bovine nuclear transfer embryos. Biol Reprod 1997; 57:204-213.
32. Baran V, Vignon X, LeBourhis D et al. Nucleolar changes in bovine nucleotransferred embryos. Biol Reprod 2002; 66:534-543.
33. Bjerregaard B, Wrenzycki Ch, Philimonenko VV et al. Regulation of ribosomal RNA synthesis during the final phases of porcine oocyte growth. Biol Reprod 2004; 70:925-935.

CHAPTER 7

Somatic Cell Nuclear Transfer (SCNT) in Mammals:
The Cytoplast and Its Reprogramming Activities

Josef Fulka, Jr.* and Helena Fulka

Abstract

I t is now more than nine years since Dolly, the world's first somatic cell cloned mammal was born, and the success of somatic cell nuclear transfer (SCNT) is still disappointingly low. Only about 3-5% of reconstructed embryos develop to term, and it is also evident that even if some clones are born, they are not necessarily fully developed and healthy. Embryonic and neonatal abnormalities of cloned offspring are probably a result of incorrect or incomplete reprogramming of the transferred donor cell nuclei. Such an incomplete reprogramming reflects the extremely low efficiency of SCNT. The key role in the process of reprogramming has been attributed to the enucleated oocyte-cytoplast into which the somatic cell nucleus is transferred. In our chapter, we will discuss the methodological approaches used for the preparation of cytoplasts and their possible reprogramming activities.

Introduction

The oocyte is the only known cell that can reprogram fully, even a terminally differentiated somatic cell nucleus in such a way that the reconstructed embryo can develop to term.[1] The technique of SCNT is fairly straightforward: the nuclear material is removed from the oocyte and a somatic cell nucleus is transferred into the enucleated oocyte-cytoplast. Although there are certain exceptions, the oocytes in metaphase II are almost exclusively used for SCNT. These oocytes can be either obtained from naturally ovulated or superovulated animals (rodents, small animals) or they are collected as immature ova from large antral ovarian follicles and then matured in vitro. The freshly isolated ovarian oocytes contain a large nucleus - germinal vesicle (GV). This nucleus disappears within a short time period when oocytes are cultured under appropriate culture conditions (mouse- 1h, cattle- 6h, pig- 18h). The oocytes then reach metaphase I stage (MI) which is followed by anaphase to telophase I transition and a subsequent meiotic arrest in metaphase II (MII; mouse 10-12h, cattle 18-24h, pig 36-40h). It is well established that oocytes produced in vitro are not of the same developmental potential as those that completed the process of maturation inside the ovarian follicles.[2,3] Moreover, it has been demonstrated in the mouse that the oocytes from superovulated animals are not as viable as those produced by natural ovulation.[4]

The fact that live animals can be born after SCNT shows that in most cases the DNA content of somatic cells (albeit their very different parameters like function, morphology and gene expression) remains the same after SCNT. This indicates that the donor cell nuclear

Corresponding Author: Josef Fulka, Jr.—Institute of Animal Production, POB 1, CS-104 01 Prague 10, Czech Republic. Email: fulka@vuzv.cz

Somatic Cell Nuclear Transfer, edited by Peter Sutovsky. ©2007 Landes Bioscience and Springer Science+Business Media.

reprogramming encompasses other mechanisms that alter neither the DNA sequence nor the DNA content. It has been shown recently that epigenetic regulations play a crucial role in both SCNT and normal development. Epigenetic inheritance is defined as heritable changes in gene expression that are not caused by alterations in the DNA sequence. DNA methylation and covalent histone modifications are among the most studied mechanisms. It is therefore probable that the state or quality of cytoplasts can dramatically influence the SCNT outcome.[5]

The eukaryotic DNA is organized into nucleosomes which are composed of an octamer of histones and approximately 146 bp long strand of DNA. The histones form a structure called the nucleosomal core particle which is composed of H2A, H2B, H3 and H4 histone molecules, each histone molecule is present in two copies. A fifth type of a histone molecule is also commonly found in eukaryotic chromatin - it is called the linker histone, or H1. The core histones can be modified by methylation, acetylation, phosphorylation and other covalent modification at many different positions (for review see ref. 6). Distinct, unique modifications can be found in different types of chromatin (euchromatin vs. heterochromatin - facultative heterochromatin vs. constitutive heterochromatin) and can also be linked to variable transcriptional states of certain genes. We will focus on histone modifications as these (especially histone acetylation) can be rather dynamic and therefore, unlike DNA or histone methylation, can be easily used to assess both the reprogramming state of somatic chromatin and the reprogramming capacity of the cytoplast.

The cytoplasts derived from metaphase II-oocytes are almost exclusively used for SCNT. However, the following section will demonstrate that the cytoplasts for SCNT can be also prepared from immature (GV, germinal vesicle stage), maturing (MI, metaphase I) or mature (MII, metaphase II) oocytes.

Immature Oocytes (GV Stage)

The immature oocytes contain a large nucleus and are typically surrounded with several layers of somatic (cumulus) cells. It must be pointed out that oocytes matured in vitro have a lower developmental potential after fertilization when compared to oocytes matured in vivo. This developmental potential is further reduced when cumulus cells are removed. The removal of cumulus cells is an essential step if these oocytes are to be manipulated (Figs. 1-4). The immature oocytes do not contain active MPF (maturation promoting factor, p34^{cdc2}/cyclin B complex) which is responsible for the initiation of maturation, chromosome condensation and meiotic progression. Thus, the nuclear envelope of somatic cells transferred into immature oocytes will remain intact for as long as the nuclear envelope of oocyte GV remains intact. Interestingly, the enlargement of the volume of transferred nuclei can be observed depending on the method used for transfer. Under the conditions used in our laboratory,[7] the nuclei introduced into immature oocytes by polyethyleneglycol (PEG) induced fusion increased their size 2 - 3 times. This increase is not evident when electric field mediated fusion is used. This volume enlargement reflects the changes occurring in somatic cell nuclei after their transfer into GV stage oocytes.

From the cell cycle perspective, the transferred nucleus must be in G2 phase. If less advanced, it must reach this stage before the oocyte cytoplasm starts to produce active MPF. It has been proposed that somatic cell chromosomes will follow the oocyte chromosome maturation pathway. After reaching metaphase II, they can be parthenogenetically activated, a process referred to as the haploidization of somatic cells.[7] If the expulsion of the second polar body is prevented by incubating the oocyte in cytochalasin B (D), the activated oocyte contains two haploid pronuclei and the resulting embryo would be diploid. Interestingly, when the randomly selected nuclei were transferred into immature oocyte cytoplasts the replication of DNA did not cease. This means that transferred nuclei completed DNA replication. The maturation of reconstructed cells was, however, very irregular. In some cases the reconstructed ova did not reach metaphase II[7] and if so, their ploidy was abnormal and an aberrant arrangement of mitotic chromosomes was observed on the oocyte meiotic spindle.

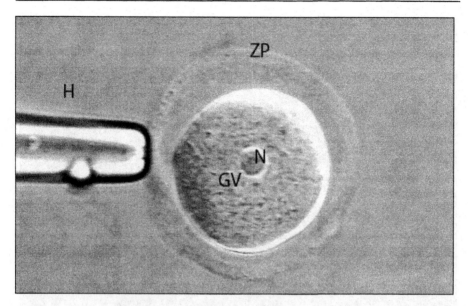

Figure 1. Enucleation of immature mouse oocytes. The oocyte is held in the holding pipette (H). ZP: zona pellucida, N: nucleolus, GV: germinal vesicle, nucleus.

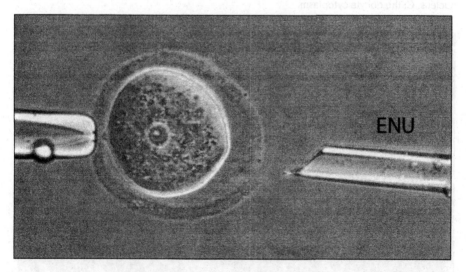

Figure 2. The enucleation pipette (ENU) that will be used for the aspiration of the oocyte nucleus.

Theoretically, the transferred nucleus could be prereprogrammed by using a transient exposure to the immature oocyte cytoplasm and then used for nuclear transfer into the conventionally prepared metaphase-II-cytoplast. As far as we are aware, this approach has never been used. Some data, however, indicate that for the induction of certain reprogramming changes, germinal vesicle material is essential. Byrne et al[8] transferred adult mammalian somatic cell nuclei from blood lymphocytes into amphibian oocyte germinal vesicles and detected their

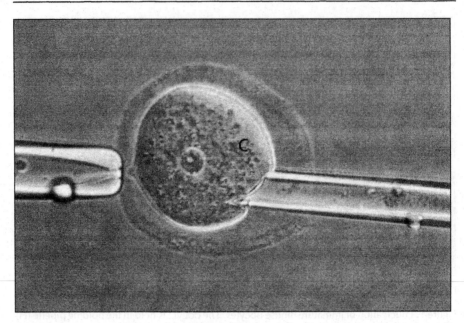

Figure 3. The enucleation pipette penetrates zona pellucida and then slowly aspirates the oocyte nucleus. C: the oocyte cytoplasm.

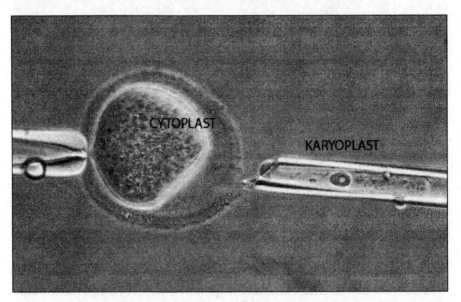

Figure 4. After the oocyte nucleus is completely aspirated into the enucleation pipette this pipette is withdrawn from the oocyte. GV (karyoplast) is then usually discarded, whilst the enucleated oocyte (cytoplast) can be used for nuclear transfer. The enucleation procedure steps are basically the same for all oocyte maturation stages.

conversion into a pluripotent state, indicated by the expression of *oct-4*. Such a conversion was not observed when the nuclei were transferred into the immature oocyte cytoplasm.

Global Histone Pattern in Somatic Cell Nuclei Transferred into Intact or Enucleated GV Stage Oocytes

It has been shown that the chromatin undergoes an extensive rearrangement during the growth phase of GV stage oocytes, and this rearrangement can be correlated to oocyte transcriptional activity.[9] It is now clear that a variety of covalent histone modifications can also be linked to either transcriptional activation or repression. For example, acetylated histones and trimethylated H3/K4 (histone H3 trimethylated on Lys 4) can be associated with active genes. On the other hand, mostly methylated histones (e.g., dimethylated H3/K9, methylated H3/K27) are found in transcriptionally repressed heterochromatin. Irrespective of their transcriptional competence, GV stage oocytes show high content of both acetylated histones at many different residues and trimethylated H3/K4 (for detailed mapping of acetylated residues see ref. 10). These findings indicate that GV stage oocytes represent a highly transcriptionally favorable environment. When a somatic cell nucleus is transferred into a cytoplast prepared from a GV stage oocyte, it often (as previously mentioned) enlarges its volume. When such reconstructed cells are examined for the presence of acetylated histones, the nuclei originating from somatic cells are labeled very strongly with antibodies against acetylated histones. Therefore, we can conclude that both the oocyte chromatin and chromatin of somatic cells transferred into these cytoplasts display marks of transcriptionally active chromatin. On the other hand, oocyte GVs also display methylation at H3/K9, which can be considered a marker of heterochromatin.

Maturing Oocytes (Metaphase I Stage)

Concomitantly with oocyte germinal vesicle breakdown (GVBD or nuclear envelope breakdown, NEBD) the chromosomes condense and are gradually arranged in metaphase I plate. This stage is followed by metaphase to anaphase I and anaphase to telophase I transition. This transition is rather short, and oocytes at this stage of meiosis are not commonly used for SCNT. The maturing oocytes contain high levels of MPF, and thus exhibit the so called "chromosome condensation activity (CCA)". When somatic cell nuclei are transferred into intact or enucleated maturing oocytes, their chromosomes typically condense under the influence of CCA. However, when the period between the oocyte enucleation and nucleus transfer is long, the transferred nuclei remain intact. Interestingly, if these nuclei replicate DNA, the replication ceases almost immediately after the transfer.[11] This means that for a correct ploidy of reconstructed embryos, the nuclei that will be eventually transferred into cytoplasts from maturing oocytes must be either in G2 or M-phase. The condensed chromosomes must then pass the anaphase to telophase I transition and extrude the first polar body (PB1). In theory, if the reconstructed oocytes are then activated and the expulsion of the second polar body is suppressed (CB, CD) the resulting embryo will be diploid.

Global Histone Pattern of Chromosomes after SCNT into Intact or Enucleated Maturing Oocytes

It has been speculated that the prolonged exposure of somatic cell chromatin to the cytoplasts factors derived from immature or maturing oocytes may improve the reprogramming of transferred nuclei and result in an improved development or reduction of abnormalities in the newborn clones. However, the results obtained to date indicate that the use of the above SCNT protocol does not produce viable embryos.[12] The reduced developmental potential of such clones could be due to the incompatibility of somatic cell nuclei and oocyte cytoplasm. This leads to an aberrant arrangement of chromosomes on the spindle and subsequently in their abnormal segregation after the exit from M-phase. We, however, cannot exclude the possibility that further modifications such as serial nucleus transfer or the induced decondenzation of transferred nuclei will overcome these obstacles.

As the oocytes enter the metaphase I, their chromatin condenses into individual chromosomes. This is accompanied by deaetylation and methylation of oocyte histones. Phosphorylation can also be detected on H3/S10 and the linker histone H1. It seems that these changes to histone modifications are rather universal and can be found both during mitosis and meiosis. When a somatic cell nucleus is transferred into a cytoplast obtained from MI mouse oocytes, its chromatin typically condenses and is rapidly deacetylated (H. Fulka, submitted). This very fast phase of epigenetic synchronization can be observed within the first hour after SCNT. Deacetylation continues beyond this initial stage and the acetylated histones are completely absent from the reconstructed cells cultured overnight (12-16h). From these observations, we can conclude that the remodeled somatic cell chromatin closely mirrors the epigenetic state of the oocyte chromatin, as shown by labeling with antibodies against various histone modifications. It is clear that cytoplasm plays a major role in the epigenetic synchronization of somatic cell chromatin, and the observed behavior is not the property of meiotic oocyte chromatin. Thus, the oocyte cytoplast is competent to induce various epigenetic changes in an exogenous chromatin derived from a somatic cell.

In certain cases, residual acetylated histones (H4/K12) can be detected on MI oocyte chromosomes (H. Fulka, submitted). When a somatic cell nucleus is transferred into such oocyte, the rapid deacetylation cannot be observed. This might indicate that a stepwise epigenetic reorganization takes place during the oocyte growth and development and the uninterrupted communication between the cytoplasm and the oocyte chromatin is absolutely essential for the achievement of a full developmental competence.

Mature Oocytes (Metaphase II)

As mentioned above, the mature MII oocytes are almost exclusively used for the preparation of cytoplasts for SCNT. This makes sense: both the ovulated oocytes and the mature, ejaculated spermatozoa are terminally differentiated cells, but soon after fertilization their genomes are merged into a single totipotent embryonic genome. Thus the oocyte reprogramming activities must be very efficient.

The most commonly used approach for the preparation of MII cytoplasts is the aspiration of metaphase II set of chromosomes, which is typically located underneath the first polar body into a beveled enucleation pipette. The aspirated portion of cytoplasm with the chromosomes is then discarded. Eventually, the enucleation procedure and its efficiency can be controlled after the oocyte chromosomes are stained with a vital fluorescent DNA stain (e.g., Hoechst33342) and then examined under UV light. This can either pinpoint the location of chromosomes in the oocyte or verify the presence of chromosomes in the aspirated oocyte portion. This step is used especially in those species (e.g., pig) where oocyte cytoplasm is too dense, first polar bodies are too small, or they rapidly disintegrate after they are extruded. In the mouse, the chromosome group can be seen in the oocyte cytoplasm as a translucent area.[13,14]

Eventually, the metaphase chromosome set can be removed by bisecting the oocyte into two portions. The smaller portion contains the chromosomes while the larger is used as a cytoplast for nuclear transfer. It must be noted, however, that such an invasive method is technically difficult and requires substantial experience and skill (handmade NT) or expensive equipment (micromanipulator, microscope, equipment for pipette preparation). Thus the simplification of the enucleation procedure is highly desirable.

In the frog, the cytoplast can be prepared relatively easily by irradiating the oocyte with UV light. When the similar approach has been tested for mammalian oocytes, no encouraging results were obtained. Promising results were obtained, however, after the irradiation of metaphase II bovine oocytes using X-ray.[15] The oocytes were exposed to 15kV(p) X-ray for different time intervals and used for SCNT. The development to blastocyst stage did not differ between conventionally enucleated and irradiated oocytes. What is more interesting is that the karyotype of blastocysts was normal after oocyte irradiation. It remains to be tested if the irradiated oocytes will support a complete development when used for SCNT.

Fulka, Jr. and Moor[16] reported that mouse oocytes can be easily and simply enucleated when exposed in metaphase I first to etoposide and thereafter to etoposide and cycloheximide. This approach led to an efficient production of fully enucleated oocyte (cytoplasts) with the entire chromosome complement expulsed from the cytoplasm and located in the first polar body, which was completely detached from the cytoplast. Moreover, the cytoplast thus produced displayed some characteristics typical for MII oocytes, including high levels of MPF and the ability to be successfully activated after SCNT. Further experiments showed that these cytoplasts can efficiently reprogram the transferred nuclei, as judged by nucleolar morphology. Unfortunately, the similar or slightly modified approaches did not allow us to enucleate the ungulate oocytes for as yet unknown reasons. The explanation could be in the asynchronous nuclear maturation, different spindle position, or in some unknown reasons.[13]

A promising approach has been developed by Overström and colleagues.[17] The oocytes in metaphase II are briefly exposed to colcemide and then parthenogenetically activated. This treatment leads to a complete expulsion of chromosome complement which seems to be completely detached from the oocyte. Several modifications were developed. In some experiments the oocytes were first exposed to colcemide and thereafter activated or vice versa. The efficiency of enucleation depends on a number of factors: the enucleation works best in the mouse, in some species—pig, rabbit—the expulsion of the polar body is not permanent so the polar body-like structure with chromosomes had to be removed by micromanipulation before SCNT. In general, such a treatment facilitates the removal of chromosomes by micromanipulation as they are located in the protrusion on the oocyte surface. More importantly, normal cloned offspring were born when nuclei were transferred into cytoplasts prepared by this type of noninvasive enucleation.

Similar to metaphase I oocyte cytoplasts, we could anticipate that the cytoplasts derived from metaphase II oocytes will also contain high levels of MPF (CCA). Consequently, the somatic cell nuclear envelope should disintegrate rapidly and chromosomes should condense. When these reconstructed embryos are then activated chromosomes will decondense, nuclei will be formed and they will eventually start to replicate DNA. Thus, nuclei transferred into MII cytoplasts should be either G0 or G1 staged to assure that the ploidy of embryos will be normal.

Global Histone Pattern in Chromosomes from Nuclei Transferred into Intact or Enucleated Mature (MII) Oocytes

Generally, we can expect that the histone modification patterns after SCNT into MII oocyte cytoplasts will be similar to histone modifications after SCNT into MI oocytes. As with MI cytoplasts, when a somatic cell is transferred into MII cytoplast, it is rapidly deacetylated within the first hour after the nuclear transfer. After an overnight incubation (12-16h), no signal can be detected with antibodies against acetylated histones (H. Fulka, submitted).

The DNA methylation after SCNT into cytoplasts prepared from MII oocytes has been described in detail and compared to zygotes from in vitro fertilization (IVF). During normal fertilization and subsequent development we can observe a phase of very extensive demethylation in many species - mouse, human, bovine.[18] It has been shown that somatic cell chromatin undergoes only limited demethylation. Beaujean et al[19] have reported that in sheep the overall methylation level in SCNT embryos stays slightly higher when compared to IVF embryos. The donor cell chromatin shows a conserved distribution when the transferred donor cell nuclei were compared to IVF embryo nuclei. In other words, the chromatin architecture or distribution in nuclei in SCNT embryos resembles the distribution in the donor cell nuclei but not in IVF embryo nuclei. Such a developmental conservation is retained in other species.[20] An aberrant methylation pattern in the mouse can be linked to a compromised embryo development.[21] Therefore, we can assume that the atypical methylation pattern observed in SCNT embryos might be a good indicator of poor developmental potential. This analysis was not performed with cytoplasts from other stages.

Activated Oocytes (Zygotes)

The original nuclear transfer protocols used the enucleated zygotes or parthenogenetically activated oocytes as a source of cytoplasts. Evidently, these cytoplasts are not convenient for SCNT. This type of cytoplast (universal, because the cell cycle stage of transferred nuclei is not important) can only be used when early, undifferentiated embryonic cells are used for NT. While the cytoplast produced by enucleation of metaphase II oocytes can be used almost universally, evident species specific differences were noted when activated oocyte cytoplasts were used, with completely negative results in the mouse and more encouraging results in ungulates. Taken together, this indicates that the oocyte reprogramming activity disappears rapidly after the oocyte activation.[22]

Global Histone Pattern in Nuclei Transferred into Intact or Enucleated Activated Oocytes

We have tested two different anti-acetylated histone antibodies on activated oocytes. These experiments revealed that the acetylation of various residues within distinct histone molecules proceeds with different dynamics. While the anti-acetyl H4/K12 signal appears very soon after the parthenogenetic pronucleus formation, H3/K9 becomes acetylated several hours later. This can be explained by the fact that during the process of oocyte maturation, the H3/K9 epitope is occupied by methylation and this excludes the possibility of acetylation at this position. Histone methylation is a very stable modification with a relatively long half-life.[23] Therefore it is likely that methyl groups must be first removed from this epitope prior to acetylation. What is, however, very interesting is the behavior of somatic cell chromatin after SCNT into activated oocytes. Here, no nuclear envelope breakdown or premature chromosome condensation can be detected. When these reconstructed embryos were investigated after labeling with anti-acetylated histone specific antibodies, it became evident that the somatic chromatin shows the distinct appearance of acetylated histones when compared to the maternal chromatin. As pointed out before, the acetylated H3/K9 can be detected in the oocyte-derived pronucleus much later when compared with acetylated H4/K12. This is, however, not true for pronuclei originating from somatic cells. In this case, both the anti acetyl H3/K9 and H4/K12 antibodies label the chromatin with approximately the same dynamics and intensity. This discrepancy can be explained by the fact that the oocyte progress through meiotic division involves the condensation of chromosomes. As previously pointed out, these events imply the methylation of histones (among other modifications). As the somatic cell is transferred into activated oocyte with no chromosome condensation occurring, it is possible that histone residues are not occupied by methylation in such an extent as in the oocyte chromatin, and therefore they can be more readily acetylated. Eventually, the original acetylation status of somatic cells stays unchanged. Thus it is highly probable that for the induction of appropriate changes the chromosome condensation is absolutely necessary.

During normal fertilization, the maternal chromosomes are transformed into a female pronucleus and the sperm chromatin into a male pronucleus. It has been shown recently that the parental pronuclei differ in the presence of many histone modifications and in the dynamics of appearance of others.[24] For example, the maternal pronucleus can be labeled by antibodies against dimethylated H3/K9 but the paternal pronucleus is negative. A different situation is observed when antibodies against hyperacetylated histones are used - both the maternal and paternal pronuclei can be visualized by these antibodies, but what differs is the dynamics of the association of parental chromatin with hyperacetylated histones, the maternal pronucleus shows a slower association.[25] Among other modifications which show a differential localization between parental pronuclei, we can also name di- and tri-methylated H3/K4.[26] From this point of view, pronuclei originating from somatic cells behave like neither the maternal nor paternal chromatin. It is possible that in order to improve SCNT technology a very sophisticated sequential transfer method must be developed which would simulate the natural changes occurring during normal meiosis and fertilization. This, however, might show as a very laborious and inefficient option.

Conclusion

The intention of this chapter was not to review comprehensively the importance of oocyte cytoplast and its ability to reprogram the transferred nuclei after SCNT. The process of donor cell nuclear remodeling and reprogramming remains complex and very poorly understood. In theory, even if the population of oocytes used for the preparation of cytoplasts is almost homogenous, and similarly the homogeneity of nuclei is very high (e.g., cumulus cells from a single oocyte), the success of SCNT remains very low. To date, the effort to increase this success yielded only an incremental improvement.

Acknowledgements

JFJr and HF are supported by grants from ESF EuroSTELLS STE/05/E004 and AVCR 0Z50390512.

References

1. Hochedlinger K, Jaenisch R. Nuclear transplantation, embryonic stem cells, and the potential for cell therapy. N Engl J Med 2003; 349(3):275-286.
2. Fulka Jr J, First NL, Moor RM. Nuclear and cytoplasmic determinants involved in the regulation of mammalian oocyte maturation. Mol Hum Reprod 1998; 4(1):41-49.
3. Moor RM, Dai Y, Lee C et al. Oocyte maturation and embryonic failure. Hum Reprod Update 1998; 4(3):223-236.
4. Hiragi T, Solter D. Reprogramming is essential in nuclear transfer. Mol Reprod Dev 2005; 70(4):417-421.
5. Li E. Chromatin modification and epigenetic reprogramming in mammalian development. Nature Rev Genet 2002; 3(9):662-673.
6. Khorasanizadeh S. The nucleosome: From genomic organization to genomic regulation. Cell 2004; 116(2):259-272.
7. Fulka Jr J, Martinez F, Tepla O et al. Somatic and embryonic cell nucleus transfer into intact and enucleated immature mouse oocytes. Hum Reprod 2002; 17(8):2160-2164.
8. Byrne JA, Simonsson S, Western PS et al. Nuclei of adult mammalian somatic cells are directly reprogrammed to oct-4 stem cell gene expression by amphibian oocytes. Curr Biol 2003; 13(14):1206-1213.
9. Christians E, Boiani M, Garagna S et al. Gene expression and chromatin organization during mouse development. Dev Biol 1999; 207(1):76-85.
10. Kim JM, Liu H, Tazaki M et al. Changes in histone acetylation during mouse oocyte meiosis. J Cell Biol 2003; 162(1):37-46.
11. Fulka Jr J, Loi P, Ledda S et al. Nucleus transfer in mammals: How the oocyte cytoplasm modifies the transferred nucleus. Theriogenology 2001; 55(6):1373-1380.
12. Gao S, Gasparrini B, McGarry et al. Germinal vesicle material is essential for nucleus remodeling after nuclear transfer. Biol Reprod 2002; 67(3):928-934.
13. Fulka Jr J, Loi P, Fulka H et al. Nucleus transfer in mammals: Noninvasive approaches for the preparation of cytoplasts. Trends Biotechnol 2004; 22(6):279-283.
14. Li GP, White KL, Bunch TD. Review of enucleation methods and procedures used in animal cloning: State of the art. Cloning Stem Cells 2004; 6(1):5-13.
15. Kim TM, Hwang WS, Shin JH et al. Development of nonmechanical enucleation method using X-ray irradiation in somatic cell nuclear transfer. Fertil Steril 2004; 82(4):963-965.
16. Fulka Jr J, Moor RM. Noninvasive chemical enucleation of mouse oocytes. Mol Reprod Dev 1993; 34(4):427-430.
17. Fisher Russell D, Ibanez E, Albertini DF et al. Activated bovine cytoplasts prepared by demecolcine-induced enucleation support development of nuclear transfer embryos in vitro. Mol Reprod Dev 2005; 72(2):161-170.
18. Santos F, Dean W. Epigenetic reprogramming during early development in mammals. Reproduction 2004; 127(6):643-651.
19. Beaujean N, Taylor J, Gardner J et al. Effect of limited DNA methylation reprogramming in the normal sheep embryo on somatic cell nuclear transfer. Biol Reprod 2004; 71(1):185-193.
20. Bour'chis D, Le Bourhis D, Patin D et al. Delayed and incomplete reprogramming of chromosome methylation patterns in bovine cloned embryos. Curr Biol 2001; 11(19):1542-1546.
21. Shi W, Haaf T. Aberrant methylation pattern at the two-cell stage as an indicator of early developmental failure. Mol Reprod Dev 2002; 63(3):329-334.

22. Campbell KHS, Alberio R. Reprogramming the genome: Role of the cell cycle. Reproduction 2003; 61(Suppl):477-494.
23. Peterson CL, Laniel MA. Histones and histone modifications. Curr Biol 2004; 14(14):R546-551.
24. Santos F, Peters AH, Otte AP et al. Dynamic chromatin modifications characterise the first cell cycle in mouse embryos. Dev Biol 2005; 280(1):225-236.
25. Morgan HD, Santos F, Green K et al. Epigenetic reprogramming in mammals. Hum Mol Genet 2005; 14(Rev Issue 1):R47-R58.
26. Lepikhov K, Walter J. Differential dynamics of histone H3 methylation at positions K4 and K9 in the mouse zygote. BMC Dev Biol 2004; 4(12):1-5.

CHAPTER 8

Mitochondrial DNA Inheritance after SCNT

Stefan Hiendleder*

Abstract

Mitochondrial biogenesis and function is under dual genetic control and requires extensive interaction between biparentally inherited nuclear genes and maternally inherited mitochondrial genes. Standard SCNT procedures deprive an oocytes' mitochondrial DNA (mtDNA) of the corresponding maternal nuclear DNA and require it to interact with an entirely foreign nucleus that is again interacting with foreign somatic mitochondria. As a result, most SCNT embryos, -fetuses, and -offspring carry somatic cell mtDNA in addition to recipient oocyte mtDNA, a condition termed heteroplasmy. It is thus evident that somatic cell mtDNA can escape the selective mechanism that targets and eliminates intraspecific sperm mitochondria in the fertilized oocyte to maintain homoplasmy. However, the factors responsible for the large intra- and interindividual differences in heteroplasmy level remain elusive. Furthermore, heteroplasmy is probably confounded with mtDNA recombination. Considering the essential roles of mitochondria in cellular metabolism, cell signalling, and programmed cell death, future experiments will need to assess the true extent and impact of unorthodox mtDNA transmission on various aspects of SCNT success.

Mitochondria and Mitochondrial DNA

Mitochondria

Mitochondria are semi-autonomous organelles that depend on both nuclear and mitochondrial genes for biogenesis and function. Mitochondria generate the bulk of cellular ATP through oxidative phosphorylation, but apart from the central role in cellular energy supply, a broad range of additional mitochondrial functions have emerged. This includes inter- and intra-cellular signaling via steroidogenesis, reactive oxygen species, nitric oxide, and calcium.[1] Moreover, mitochondria and their components are crucial to at least three signaling pathways in programmed cell death and may affect embryonic and fetal development via apoptotic pathways.[2] Additional functions have also been assigned to specific mitochondrial gene products (see "Mitochondrial DNA" section). At least three of the mtDNA encoded respiratory chain proteins (NADH dehydrogenase 1, cytochrome c oxidase 1, and ATP synthase 6) are processed to yield maternally transmitted antigens in rodents.[3] The mitochondrial 12S and 16S rRNAs are involved in protein folding and may function as molecular chaperones.[4]

Mitochondrial distribution, morphology, and activity are characteristic for specific developmental stages in oocytes and preimplantation embryos and correct spatial and temporal changes are important determinants of successful embryonic development.[1] Each blastomere is entirely dependent on the energy produced from the portion of mitochondria that

*Stefan Hiendleder—JS Davies Fellow, Department of Animal Science, The University of Adelaide, Roseworthy Campus, Roseworthy, South Australia 5371, Australia. Email: Stefan.Hiendleder@adelaide.edu.au

Somatic Cell Nuclear Transfer, edited by Peter Sutovsky. ©2007 Landes Bioscience and Springer Science+Business Media.

was inherited from the oocyte, and reduced competence may occur during embryo cleavage if blastomeres inherit a mitochondrial complement with insufficient ATP generating capacity.[5] Notably, ATP levels in SCNT mouse embryos are much more variable than in controls, a profile consistent with representation of cleavage stage embryos as mosaics, whereby intra-embryo variations compound inter-embryo variations.[6]

Mitochondria have long been depicted as a more static population of individual organelles, with cell type dependent but mostly spheroid or tubular shape, because standard electron micrographs showed only single cross sections of the cell. However, time-lapse microscopy data have now clearly demonstrated that mitochondrial morphology is cell cycle dependent and highly dynamic, with rapid changes within a cell. This is caused by constant fusion and fission events that establish mitochondrial networks. Mitochondria thus represent a single cellular compartment, a reticulum. Fusion promotes intermitochondrial cooperation and fission enables compartmentalization. Experimental perturbation of mitochondrial fusion causes loss of mtDNA stability, defects in mitochondrial respiration, poor cell growth with increased susceptibility to cell death, and results in embryonic malformations and mortality. This demonstrates the central role of mitochondrial dynamics and structure for the dissipation of energy, signal transmission and apoptosis, cell developmental processes, and the inheritance of mtDNA.[7-9]

Mitochondrial DNA

Mitochondrial DNA (mtDNA) is a circular molecule of approximately 16-17 kb. The gene content and organization of mammalian mtDNA were determined in the first human genome sequencing project that was completed in 1981.[10] Currently, complete mtDNA reference sequences are available from publicly accessible databases for all species where SCNT has been successfully applied. This includes nonmammalian species such as xenopus (*Xenopus laevis*) and zebrafish (*Danio rerio*).[11] The mutation rate in mtDNA is apparently driven by reactive oxygen species and much higher than in nuclear DNA, which has lead to extensive intraspecific sequence variation.[12] In many domestic animal species that are subject to SCNT, considerable mtDNA haplotype variation has been incorporated from different wild ancestral subspecies.[13]

The mtDNA molecule encodes for 37 genes and contains a single major noncoding control region (CR). The CR contains regulatory sequences for transcription and replication intitiation, including a triple-stranded structure known as displacement (D)-loop. Two rRNA and 22 tRNA genes contribute to the mitochondrial translation apparatus. The remaining 13 genes encode essential subunits of enzymes involved in cellular ATP production by oxidative phosphorylation. This includes 7 subunits of respiratory chain complex I (NADH dehydrogenase), the cytochrome *b* in complex III (ubiquinol:cytochrome c oxidoreductase or bc$_1$-complex), three subunits of complex IV (cytochrome c oxidase) and two subunits of complex V (ATP synthase).[14] All other proteins involved in mitochondrial biogenesis and functions, currently estimated at ~1,500, are encoded by nuclear DNA, synthesized in the cytosol, and are targeted and imported into mitochondria by specific mechanisms. Extensive nuclear-mitochondrial interactions for the coordination of cellular signaling frameworks and transcriptional regulatory circuits are thus required to ensure proper mitochondrial biogenesis and function.[15]

Mitochondrial DNA is packaged and organized in protein-DNA complexes termed mitochondrial nucleoids. In human cells, each nucleoid contains from 2-20 mtDNA molecules. Some of the identified nucleoid proteins, including mitochondrial transcription factor A (TFAM), have known functions in mtDNA replication and transcription. However, the known functions of many other proteins are ostensibly unrelated to mtDNA function. It appears that mitochondria have recruited various proteins of mitochondrial metabolism and biogenesis to mt-nucleoids to organize and protect mtDNA, drive the segregation of the organellar genome, and couple the inheritance of mtDNA with cellular metabolism. This suggests that mtDNA acts as a central hub that organizes its immediate surroundings both within, and outside, the mitochondrion to integrate mitochondrial and cellular protein synthesis.[9,16]

Mitochondrial DNA Inheritance

Mitochondrial DNA Inheritance in Natural Reproduction

Maternal inheritance of mammalian mtDNA was first described in horse × donkey hybrids by Hutchison et al in 1974.[17] Similar low resolution data were later reported by others for rodents and human. Maternal inheritance was attributed to the massive difference in mtDNA copy number between oocytes and spermatozoa as determined by quantitative hybridization experiments.[18] More recently, real-time quantitative PCR has detected ~ 10 mtDNA copies in spermatozoa.[19] Corresponding estimates for oocytes range from 317 - 795 × 10[5] mtDNA copies, with a very high inter-oocyte variation.[20-22]

Paternally inherited mtDNA molecules were first detected by PCR in *Mus musculus* × *M. spretus* hybrids after numerous generations of backcrossing. The presence of paternal mtDNA was attributed to a selective advantage of paternal mtDNA in a compatible nuclear background and/or a mechanism that recognized and eliminated only intraspecific mtDNA.[23] The comparative analysis of *M. musculus* intraspecies hybrids and *M. musculus* × *M. spretus* interspecies hybrids confirmed the latter hypothesis. In intraspecies hybrids, paternal mtDNA could only be detected through the early pronucleus stage. In the interspecies hybrids, however, paternal mtDNA was detected throughout development to neonates. An additional experiment showed that sperm mitochondria from a conplastic strain, which carried *M. spretus* mtDNA on a background of *M. musculus* nuclear genes, were eliminated by *M. musculus* oocytes. Kaneda et al[24] therefore concluded that the species-specific exclusion mechanism recognized proteins in the sperm midpiece and not the mtDNA itself. Subsequent research data by Sutovsky et al[25,26] have clearly shown that the exclusion mechanism for sperm mitochondria consists of an ubiquitin tag, which causes proteasomal and/or lysosomal degradation of mitochondria after fertilization.

The dogma of exclusively maternal intraspecific mtDNA inheritance was shaken in 2002 by the first well-documented case of paternal mtDNA in muscle tissue of a heteroplasmic patient suffering from mitochondrial myopathy.[27] Heteroplasmic tissue from the same patient was subsequently used to demonstrate recombination between the maternal and paternal mtDNA haplotypes and thus overturned a second mtDNA dogma.[28] The analysis of additional heteroplasmic individuals has since indicated that mtDNA recombination is common in human skeletal muscle.[29]

Mitochondrial DNA Inheritance after SCNT

Mitochondrial DNA Inheritance after Intraspecies SCNT

The analysis of mtDNA in blastomere cloned embryos and offspring revealed varying degrees of heteroplasmy that could be attributed to mtDNA from the nuclear donor cell. It was thus evident that mtDNA derived from sources other than spermatozoa could escape the selective mechanism that targets and eliminates intraspecific sperm mitochondria in the oocyte to maintain homoplasmy. These findings were in line with reports of mtDNA transmission after ooplasmic transfer in human and after microinjection of somatic mitochondria into mouse zygotes, which caused dose dependent heteroplasmy levels.[12]

Surprisingly, the first investigation on mtDNA inheritance in animals generated by SCNT failed to detect nuclear donor cell mtDNA. None of the investigated sheep (*Ovis aries, O. musimon*),* including Dolly, showed evidence for heteroplasmy at detection limits of 0.5 and 0.1%.[30,31] However, subsequent mtDNA studies in bovine (*Bos taurus, B. indicus*),[32-38] goat (*Capra ibex, C. hircus*),[39] porcine (*Sus scrofa, S. vitattus*),[40,41] and mouse (*Mus musculus domesticus,*

---

*Please note that contrary to current nomenclature *Ovis aries* and *O. musimon, Bos taurus* and *B. indicus*, and strains and wild progenitors of other domestic animals are conspecific (see "Mitochondrial DNA Inheritance after Intraspecies SCNT" for further details).

M. m. molossinus)[42] SCNT embryos, fetuses and offspring identified varying levels of nuclear donor mtDNA (Table 1). The available data are still limited and often difficult to interpret because of the different experimental procedures involved in the generation of SCNT embryos and offspring subjected to mtDNA analyses. Nuclear donor cell type, passage number of donor cells and culture conditions of embryos are of particular relevance in this respect. Moreover, some analytical protocols and biological materials used in the assessment of heteroplasmy were not evaluated with respect to confounding variables (see "Confounding in Mitochondrial DNA Analysis" section). Nevertheless, some general conclusions can be drawn from the currently available data. (i) The majority of investigated ovine, bovine and porcine SCNT individuals appear to be homoplasmic or display only mild heteroplasmy that is consistent with neutral segregation of nuclear donor cell mtDNA.[30-41] (ii) The degree of heteroplasmy is tissue-specific.[32,34-36,41,42] (iii) Some SCNT individuals show more pronounced or even extreme levels of heteroplasmy that are inconsistent with neutral segregation of mtDNA.[34,36,41,42] The higher levels of heteroplasmy appear to be more frequent in intrasubspecific combinations of donor cells and recipient oocytes[34,41,42] and might thus be related to interactions with specific nuclear genes that control mtDNA segregation.[44] It is of particular note that female offspring from heteroplasmic SCNT animals can transmit heteroplasmy to the next generation after natural breeding.[41]

Mitochondrial DNA Inheritance after Interspecies SCNT

Interspecies SCNT (iSCNT) experiments create xenomitochondrial hybrids and are performed to address basic research questions, to make use of readily available oocytes of related domestic species for the preservation of endangered species, and to explore the possibility of a substitution of human oocytes in stem cell production (i.e., therapeutic cloning). Considering the essential functions of mitochondria (see "Mitochondria and Mitochondrial DNA"), data on mtDNA inheritance after iSCNT is of particular interest and deserves detailed analysis and discussion. The identification of true iSCNT is vital to the conclusions drawn from this type of experiment. Considerable confusion has been created by current zoological nomenclature that assigns species-status to populations that are often subspecies at the most. This is certainly the case for many domestic strains of animals and their wild ancestors. Examples are *Bos taurus* and *B. indicus* cattle,[45] *Ovis aries* and *O. musimon* (also referred to as *O. aries musimon*) including other wild sheep,[13] *Capra hircus* and *Capra ibex* goat,[46] and *Sus scrofa* and *S. vitattus* strains of domestic pig.[47] For information on mtDNA inheritance after SCNT in these species see Table 1.

From the condensed iSCNT data in Table 2 it is obvious that heteroplasmy was detectable in all but one of the very diverse iSCNT combinations under study.[48-56] However, the gaur/cattle fetuses that showed homoplasmy in various tissues at 1% detection limit represent the only species-combination that can crossbreed and give birth to viable offspring.[1] Heteroplasmy was thus present in preimplantation stages of all iSCNT combinations of species that can not give rise to viable offspring by natural reproduction.[49-56] The present iSCNT data are biased for rabbit oocytes, and the number of studies that employed a quantitative approach are limited. It is nevertheless obvious that all quantitative studies detected a pronounced reduction of nuclear donor cell mtDNA after the 16-cell (*Capra ibex/Oryctolagus cunniculus*,[54] *Homo sapiens/Bos taurus*[55]) or morula (*Macaca mulatta/O. cunniculus*,[51] *Bos taurus/O. cunniculus*[53]) stage. Changes in heteroplasmy levels could thus be a consequence of major genome activation in embryos that can be delayed in iSCNT combinations.[1] This result is surprising, since conspecific somatic nuclear donor cell mtDNA could be expected to show a replicative advantage over a foreign species of oocyte mtDNA. Sansinena et al[57] have recently described that foreign oocyte mitochondria that were introduced into the recipient oocyte by ooplasmic transfer did not mix with recipient oocyte mitochondria in the developing embryo, but remained in a distinct cluster even after 144 h of in vitro culture. This could also apply to foreign nuclear donor cell mtDNA in iSCNT. In addition, mouse-rat and human-nonhuman primate cybrid

Table 1. *Levels of heteroplasmy detected in embryos and offspring derived from intraspecies somatic cell nuclear transfer*

Species[j] and Reference	Offspring Analyzed	Degrees of Heteroplasmy Observed (% Nuclear Donor Cell mtDNA)	Detection Methods (Assay Sensitivity[ii])
Sheep (*Ovis aries*) [Evans et al, 1999[30]]	10 sheep of unspecified age derived from3 nuclear donor cell types	No heteroplasmy in blood, milk, muscle, and placenta, 1-7 sampled individuals per tissue.	Southern blot hybridization with phosphorimager (0.1%); cloned PCR fragments (0.5%)
Sheep (*Ovis aries musimon* and *O. aries*) [Loi et al[31]]	1 lamb derived from a mouflon somatic cell transferred into domestic sheep recipient oocyte	No heteroplasmy in blood	Direct sequencing of PCR products from SCNT lamb, oocyte donor and recipient sheep (not specified)
Cattle (*Bos taurus*) [Steinborn et al, 2000[32]]	6 fetuses (day 60-212) and 4 calves derived from 3 nuclear donor cell types	0.1-4% heteroplasmy in all but one sample in blood, cerebellum, heart, kidney, liver, lung, muscle, and skin with 1-5 sampled individuals per tissue.	Allele-specific real-time PCR (0.1%)
Cattle (*Bos taurus*) [Do et al, 2002[33]]	1-, 2-, 4-, 8-, 16- cell, morula and blastocyst stage embryos (numbers not specified) derived from one nuclear donor cell type	Heteroplasmy was detected by allele-specific PCR at all developmental stages of embryos and was confirmed by direct sequencing of allele-specific PCR product in 2 of 5 blastocysts analyzed in this manner	Allele-specific PCR on agarose gel (not specified) and direct sequencing of PCR products (not specified)
Cattle (*Bos taurus, B. indicus*) [Steinborn et al, 2002[34]]	20 cows derived from a single nuclear donor cell type with *B. taurus* nucleus and *B. indicus* mtDNA; 11 and 9 cows, respectively, were generated from oocytes with *B. taurus* and *B. indicus* mtDNA	Varying levels of heteroplasmy in all samples but oocytes (n = 2); heteroplasmy in blood (n = 19): 0-2.8%, follicular cells (n = 9): 0-12.7%, skin (n = 9): 0-2.1%, muscle (n = 8): 0-7.3%; higher levels of heteroplasmy were detected in cows generated from oocytes with *B. indicus* mtDNA, but data are confounded with passage number of nuclear donor cells and embryo culture protocol	Allele -specific real-time PCR (0.1% and 0.5 % in cows generated from oocytes with *B. taurus* and *B. indicus* mtDNA, respectively)

Table continued on next page

Table 1. *Continued*

Species[j] and Reference	Offspring Analyzed	Degrees of Heteroplasmy Observed (% Nuclear Donor Cell mtDNA)	Detection Methods (Assay Sensitivity[ji])
Cattle (*Bos taurus, B. indicus*) [Hiendleder et al, 2003[35]]	12 day 80 fetuses derived from one nuclear donor cell type and 3 different types of recipient oocyte cytoplasms (2 *B. taurus*, 1 *B. indicus*)	No heteroplasmy at 2% assay sensitivity in blood, brain, heart, jejunum, kidney, liver, lung, muscle, rumen, placenta, skin, and spleen with 12 sampled individuals per tissue. 0.5-0.7% heteroplasmy in brain, jejunum, liver, muscle, and placenta at 0.5% assay sensitivity; recipient cow derived heteroplasmy of 2.5% and 5% in 2 fetal blood samples	PCR-RFLP on agarose gel (2%), cloned PCR fragments (0.5%)
Cattle (*Bos taurus*) [Takeda et al 2003[36]]	16 embryos derived from 3 nuclear donor cell types, 1 type with or without serum starvation, analyzed directly after fusion; 61 calves and fetuses (age and numbers not specified) derived from 9 nuclear donor cell types	No heteroplasmy in fused embryos by SSCP analysis, but heteroplasmy detected in 3 embryos by Southern hybridization; 4 calves showed heteroplasmy not due to the nuclear donor mtDNA, 4 calves showed 'low' amounts of nuclear donor mtDNA in SSCP analysis, 3 calves showed massive tissue-specific heteroplasmy for nuclear donor mtDNA (0-59%) in brain, heart, kidney, liver, lung, muscle, ovary, skin, spleen, thymus, and thyroid	SSCP (3-4%) and Southern blot hybridization (not specified)
Cattle (*Bos taurus*) [Han et al 2004[37]]	5 calves derived from one nuclear donor cell type	Heteroplasmy detected in skin samples of all 5 calves and confirmed by sequencing of allele-specific products	Allele-specific PCR on agarose gel (not specified)
Cattle (*Bos taurus*) [Theoret et al, 2005[38]]	3 calves derived from one nuclear donor cell type	Heteroplasmy detected in skin (0.7-2.3%) and blood (1-3.6%) samples of all 3 animals; note that the marker for donor cell mtDNA is a length variant caused by a tandemly repeated 21 bp segment and thus might not be a stable polymorphism[12]	Agarose gel electrophoresis and image analysis (not specified)
Goat (*Capra ibex and C. hircus*) [Jiang et al, 2004[39]]	1-, 2-, 4-cell, and morula, stage embryos derived from *C. ibex* nuclear donor cells and *C. hircus* recipient oocytes (numbers not specified)	Heteroplasmy detected in 1- and 2-cell stage embryos but *C. ibex* mtDNA undetectable in morulae	Allele-specific PCR on agarose gel (not specified) and sequencing of PCR - product

Table continued on next page

Table 1. Continued

Species[i] and Reference	Offspring Analyzed	Degrees of Heteroplasmy Observed (% Nuclear Donor Cell mtDNA)	Detection Methods (Assay Sensitivity[ii])
Pig (*Sus scrofa*) [St John et al, 2005[40]]	5 pigs (generated by double nuclear transfer[43]) derived from 2 nuclear donor cell types	No evidence for donor cell mtDNA in blood samples but low levels of other heteroplasmic variants attributed to bimaternal inheritance; note that in double nuclear transfer, the donor cell is initially transferred into an enucleated recipient oocyte, and is then, as a pseudo pro-nucleus, transferred to an enucleated recipient zygote	Allele-specific real-time PCR (not specified). Primer extension polymorphism (< 0.1%)
Pig (*Sus scrofa, S. vitattus*) [Takeda et al, 2006[41]]	4 SCNT pigs (F_0) derived from *S. vitattus* nuclear donor cells and *S. scrofa* recipient oocytes; 25 progeny (F_1) of SCNT sows generated by natural mating; 18 next generation (F_2) offspring from a single F_1 sow	0.1-1% heteroplasmy in blood and hair roots of all F_0 pigs; 1 of 24 F_1-progeny showed nuclear donor cell derived heteroplasmy in blood (0.1-1%), heart (25%), kidney (4.3%), liver (44%), lung (6%), and muscle (12%); 18 F_2-offspring of the heteroplasmic F_1-female were again heteroplasmic in blood (5.8 ± 3.7%), ear (6.7 ± 5.3%), liver (12.9 ± 8.3%), and spleen (5.0 ± 3.9%)	SSCP (5%), agarose gel PCR-RFLP with image analysis (1%), and allele-specific PCR (0.1%)
Mouse (*Mus musculus domesticus, M. m. molossinus*) [Inoue et al, 2004[42]]	25 adult mice (day 115-403) derived from 3 nuclear donor cell types (cumulus-, sertoli-, and fibroblast cells) with *M. m. molossinus* mtDNA and oocytes with *M. m. domesticus* mtDNA	Heteroplasmy in brain, kidney, liver, and tail of all but one animal; the amount of nuclear donor cell mtDNA depended on donor cell type and tissue with significant differences between sertoli (0.78 ± 0.19%) and fibroblast (2.4 ± 0.7%) derived mice, and between brain (0.6 ± 0.1%) and liver (1.7 ± 0.5%) tissues; heteroplasmy detected in all 20 placentae studied (cumulus cells: 0.15 ± 0.10%, sertoli cells: 0.24 ± 0.02%, fibroblast cells: 0.54 ± 0.24%)	Allele-specific real-time PCR (not specified)

[i] Listed experiments refer to intraspecies and intrasubspecies SCNT. See text for details; [ii] As listed by the authors

Table 2. Levels of heteroplasmy detected in embryos and offspring derived from interspecies somatic cell nuclear transfer (iSCNT)

Nuclear Donor/ Recipient Oocyte Species [Reference]	Offspring Analyzed	Degrees of Heteroplasmy Observed (% Nuclear Donor Cell mtDNA)	Detection Methods (Assay Sensitivity)
Gaur (*Bos gaurus*)/ cattle (*B. taurus*) [Lanza et al, 2000[48]]	3 fetuses day 46-54 derived from 1 nuclear donor cell type	No heteroplasmy in brain, eye, gonad, heart, intestine, kidney, liver, lung, muscle, skin, and tongue	PCR-RFLP and Phosphorimager (1%); Allele-specific PCR and Phosphorimager (1%)
Giant panda (*Ailuropoda melanoleuca*)/ rabbit (*Oryctolagus cuniculus*) [Chen et al, 2002[49]]	Blastocysts (number not specified) and 2 fetuses recovered from a recipient cat uterus 21 days after transfer	Heteroplasmy detected in blastocysts; both fetuses displayed only giant panda mtDNA	Species-specific PCR on agarose gel (not specified); sequencing of PCR-product
Macaque (*Macaca mulatta*/rabbit (*Oryctolagus cuniculus*) [Yang et al, 2003[50]]	1-, 2-, 4-, 8-, 16-cell, morula and blastocyst stage embryos (numbers not specified) derived from one nuclear donor cell type	Heteroplasmy detected in all developmental stages of preimplantation embryos and confirmed by sequencing in one 4-cell embryo and one blastocyst	Species-specific PCR on agarose gel (not specified); direct sequencing of PCR products
Macaque (*Macaca mulatta*/rabbit (*Oryctolagus cuniculus*) [Yang et al, 2004[51]]	1-, 2-, 4-, 8-, 16-cell, morula and blastocyst stage embryos derived from one nuclear donor cell type; 5 embryos per developmental stage pooled for triplicate analyses	Heteroplasmy detected in all developmental stages of preimplantation embryos; the ratio of macaque mtDNA/rabbit mtDNA was $2.0 \pm 0.8\%$ (1-cell), $1.3 \pm 0.6\%$ (2-cell), $1.7 \pm 1.3\%$ (4cell), $0.6 \pm 0.3\%$ (8-cell), $0.8 \pm 0.2\%$ (16-cell), $1.3 \pm 0.5\%$ (morulae), and $0.011 \pm 0.007\%$ (blastocysts)	Species-specific real-time PCR (standard curves from $10 - 10^6$ and $10 - 10^7$ copies for macaca and rabbit mtDNA, respectively); Sequenced PCR products
Human (*Homo sapiens*)/ rabbit (*Oryctolagus cuniculus*) [Chen et al, 2003[52]]	Blastocysts derived from different nuclear donor cell types (numbers not specified)	Heteroplasmy detected in blastocysts; stem cells derived from such blastocysts appeared also heteroplasmic (data not shown)	In situ hybridization with fluorescence-labeled probes specific for rabbit mtDNA (not specified); cross-species checked PCR

Table continued on next page

Table 2. *Continued*

Nuclear Donor/ Recipient Oocyte Species [Reference]	Offspring Analyzed	Degrees of Heteroplasmy Observed (% Nuclear Donor Cell mtDNA)	Detection Methods (Assay Sensitivity[i])
Cattle (*Bos taurus*)/ rabbit (*Oryctolagus cunniculus*) [Jiang et al, 2006[53]]	1-, 2-, 4/8-, morula and blastocyst stage embryos derived from 2 nuclear donor cell types (from a heteroplasmic SCNT animal and its homoplasmic nuclear donor); at least 6 individual embryos per developmental stage analyzed	Heteroplasmy detected in all developmental stages of preimplantation embryos; the amount of rabbit mtDNA remained constant to the morula stage and was increased sharply in both types of blastocysts; bovine mtDNA remained at constant levels in the SCNT nuclear donor cell group during preimplantation development but decreased sharply in the non SCNT nuclear donor group after the morula stage	Species-specific real-time PCR (standard curves from 10^2 - 10^6 and 10 - 10^5 mtDNA copies for rabbit and bovine, respectively); Cross-species checked PCR
Ibex (*Capra ibex*)/rabbit (*Oryctolagus cunniculus*) [Jiang et al, 2006[54]]	2-, 4-, 8-, 16-cell, morula and blastocyst stage embryos derived from one nuclear donor cell type (numbers not specified)	Heteroplasmy detected in all developmental stages of preimplantation embryos; rabbit mtDNA increased from the 16-cell stage on, ibex mtDNA decreased after the 16 cell-stage	Species-specific PCR on agarose gel (not specified); 30 cycles for rabbit, 40 cycles for Ibex mtDNA; cross-species checked PCR; sequencing of cloned PCR fragment
Human (*Homo sapiens*)/ bovine (*Bos taurus*) [Chang et al, 2003[55]]	2-, 4-, 8-, 16-cell, morula and blastocyst stage embryos derived from 5-2 nuclear donor cell types (numbers not specified, but 5-7 replicates indicated)	Heteroplasmy detected in all 2-, 4-, 8-, 16-cell stage embryos; only *B. taurus* mtDNA detected after 16-cell stage	Species-specific PCR on agarose gel with image analysis (not specified); cross-species checked PCR; sequenced cloned PCR fragment
Takin (*Budorcas taxicolor*)/ Yak (*Bos grunniens*) [Li et al, 2006[56]]	2 blastocysts derived from a single nuclear donor cell type	Heteroplasmy detected in both blastocysts	Species-specific nested PCR on agarose gel (not specified); sequencing of PCR-products

[i] As listed by the authors

cells have demonstrated that foreign mtDNA is replicated in spite of the extensive nuclear-mitochondrial incompatibility that results in mitochondrial dysfunction.[1] This might explain the heteroplasmy observed in blastocysts and stem-cells derived from human donor cells and rabbit oocytes[52] but can not account for the exclusive presence of giant panda (Ailuropoda melanoleuca) mtDNA in two fetuses generated from panda nuclear donor cells and rabbit (Oryctolagus cunniculus) oocytes.[49] Clearly, more data are needed, and potentially confounding effects (see below), including parthenogenesis, have to be controlled.

Confounding in Mitochondrial DNA Analysis

The analysis of mtDNA inheritance and segregation often requires the detection and quantification of comparatively low amounts of a minor mtDNA haplotype in a high amount of a second, major mtDNA haplotype. This is typically the case when somatic nuclear donor cell mtDNA is quantified in a background of recipient oocyte mtDNA where the initial ratio of donor cell to oocyte mtDNA copy number is < 1:100.[42] Therefore, confounding true organelle mtDNA with nuclear mitochondrial pseudogenes (numts) poses a serious problem in PCR and hybridization based methods. Nuclear integration of mtDNA sequences is a common and continuous process, and hundreds of numts have been identified in various species.[58] Because of the high number of integration sites, and the high sequence identity of recently integrated numts to their mitochondrial counterparts, numts may readily coamplify with mtDNA.[59] Preferential amplification of numts has also been observed, especially when PCR primers based on conserved mtDNA sequences from related taxa were used.[60]

Natural heteroplasmy is another potentially confounding variable. Heteroplasmy caused by tandem-repeat sequences in the mtDNA CR has been observed in numerous mammalian taxa, and repeat regions should be avoided in primer design.[61] Heteroplasmy resulting from single nucleotide polymorphism or contraction/expansion of short mononucleotide repeats is of far greater concern. Examples include the mutation hot spots in hypervariable segments of the mtDNA CR.[62] This region is often targeted in mtDNA analyses because of its high intraspecific sequence variation. Hypervariable sites that mutate spontaneously can display marked differences in their heteroplasmic ratios, and these ratios may differ in an organ-specific manner. Single strand conformation polymorphism (SSCP) analyses are particularly prone to pick up such events, but the analysis of single restriction sites with hypervariable nucleotide positions can also yield erroneous results.

A further possible complication in mtDNA analysis is chimerism. Blood lymphocyte and germ cell chimerism that results from vascular anastomoses and reciprocal exchange of cells between embryos during intrauterine development is a well-known phenomenon in bovine twinning. But this type of chimerism has also been observed in other mammals, including horse, sheep, human, and lama.[12] In bovine twin fetuses that were generated from maternally unrelated oocytes by in vitro fertilization and cotransferred to the same recipient cow as embryos, chimerism was readily detected in blood and liver samples by PCR-RFLP analyses of mtDNA.[63] Moreover, recipient cow mtDNA has been detected in blood samples of SCNT fetuses. It is at present not clear if this phenomenon is related to the slaughtering procedure or due to abnormal placentation in SCNT pregnancies. Maternal cells appear to frequently cross the placental barrier in utero in human and during parturition in bovine.[63]

In order to avoid the pitfalls in mtDNA analysis, studies on mtDNA transmission and segregation after SCNT should be based on experimental designs with prior knowledge of all involved mtDNA haplotypes.[35] Primer combinations located in different segments of the mitochondrial genome, testing for coamplification of numt sequences on mtDNA-less ρ^{o} cells or sperm heads, and sequencing of PCR products can further reduce the chance of confounding data. The assessment of mtDNA recombination in heteroplasmic SCNT individuals will have to take the possibility of jumping PCR artifacts (i.e., hybrid molecules generated by template switching) into account.

Implications of SCNT Mediated Mitochondrial DNA Inheritance

A survey of perinatal clinical data from human subjects with deficient mitochondrial respiratory chain activity has revealed phenotypes that have striking similarities with abnormalities encountered in SCNT fetuses and offspring.[1] Nuclear and mitochondrial genes have co-evolved to ensure the very tight and precise interactions that are required for proper mitochondrial biogenesis and function. This is also responsible for a species-specific incompatibility between nuclear and mitochondrial genomes. Minor mutations in mtDNA or in nuclear genes involved in mitochondrial interactions can lead to severe diseases.[14] Standard SCNT procedures inevitably generate a constellation that would never occur in nature, i.e., deprive an oocytes' mtDNA of the corresponding maternal nuclear DNA and force it to interact with an entirely foreign nucleus that is again interacting with foreign somatic mitochondria.

Murine-murine, murine-rat and human-nonhuman primate cybrid cells have demonstrated various degrees of nuclear-mitochondrial incompatibility that can result in mitochondrial dysfunction with impaired oxygen consumption, deficiency in respiratory chain enzymes, reduced cell growth, increased free radical production, increased lipid peroxidation, loss of membrane potential, and apoptosis.[1] Similar problems with mitochondrial phenotype could be expected in cells derived from iSCNT, but nuclear-mitochondrial incompatibility might also impact on intraspecies SCNT.

In natural reproduction, at least two mechanisms operate to maintain or restore homoplasmy. The intraspecific exclusion mechanism that prevents paternal leakage of mtDNA (see "Mitochondrial DNA Inheritance in Natural Reproduction" section) is complemented by a severe bottleneck in mtDNA segregation during oogenesis. mtDNA is susceptible to mutation because of exposure to high concentrations of reactive oxygen species in mitochondria. The bottleneck reduces the number of segregating mtDNA molecules to < 10 and is thought to operate as a selection mechanism to maintain mitochondrial genomic integrity because of its critical role in cellular metabolism and functions.[64] Homoplasmy is therefore expected to be required for proper mitochondrial function.

Heteroplasmy could be responsible for the high death rates associated with nuclear transfer,[25] a notion supported by the reduced survival of parthenogenetic embryos after injection of somatic cytoplasm or mitochondria.[65] This could explain the observed homoplasmic or nearly homoplasmic state in the majority of analyzed individuals derived from SCNT. Data on mtDNA heteroplasmy in dead and aborted fetuses is extremely limited but have so far not revealed extensive heteroplasmic conditions.[12] The cases of clearly documented more extreme and tissue-specific heteroplasmy nevertheless raise the possibility of a more frequent initial occurrence of heteroplasmy with detrimental effects on embryo development and subsequent embryonic loss. Interestingly, the most extensive experiment to assess heteroplasmy in intraspecies SCNT has revealed a nuclear donor cell type and/or SCNT procedure dependent degree of heteroplasmy. Electrofusion of fibroblasts caused higher levels of donor cell mtDNA to be present in offspring than cumulus or sertoli cell microinjection. This established a link between nuclear donor cell type, SCNT procedure, mtDNA heteroplasmy, and SCNT success.[42]

Extreme heteroplasmy is expected in embryos and offspring generated by "hand made somatic cell cloning" where oocytes are split and two enucleated demi-oocytes are combined with the somatic nuclear donor cell in a two step fusion process. Reconstructed embryos are therefore expected to be triplasmic, i.e., to contain three types of mitochondria: oocyte mitochondria from two different recipient cytoplasts and the mitochondria from the nuclear donor cell.[1] "Hand made somatic cell cloning" would therefore aggravate potential problems caused by heteroplasmy. Interestingly, a recent study showed that the transfer of "hand made" bovine SCNT embryos to recipients resulted in an initially very high pregnancy rate followed by an unusually high intrauterine mortality.[66]

An important point to consider in iSCNT directed towards therapeutic cloning is the observed function of mitochondrial genes as minor cell surface antigens.[3] A "proof of principle" experiment in the cow (*Bos taurus*) demonstrated that bioengineered tissues from SCNT showed long-term viability after transplantation of the grafts into the nuclear donor animals despite expressing a different mtDNA haplotype. However, mtDNA variation in this experiment was low and would certainly be much higher in cells from iSCNT (see "Mitochondrial DNA Inheritance after Inter-Species SCNT" section).

Many of the potential problems associated with mtDNA inheritance in SCNT could be overcome by using somatic nuclear donor cells without mitochondria, an approach that is currently under investigation.

Conclusion

Both intraspecies and interspecies SCNT leads to unorthodox mtDNA transmission and inheritance. The consequences of identified and predicted perturbances of the normal homoplasmic state are at present unknown. Experiments designed to take the pitfalls in mtDNA analysis into account are needed to assess the true extent and potential impact of nuclear-cytoplasmic incompatibility, heteroplasmy, and mtDNA recombination on the various aspects of SCNT success.

References

1. Hiendleder S, Zakhartchenko V, Wolf E. Mitochondria and the success of somatic cell nuclear transfer cloning: From nuclear-mitochondrial interactions to mitochondrial complementation and mitochondrial DNA recombination. Reprod Fertil Dev 2005; 17:69-83.
2. Mirkes PE. 2001 Warkany lecture: To die or not to die, the role of apoptosis in normal and abnormal mammalian development. Teratology 2002; 65:228-239.
3. Bhuyan PK, Young LL, Lindahl KF et al. Identification of the rat maternally transmitted minor histocompatibility antigen. J Immunol 1997; 158:3753-3760.
4. Sulijoadikusumo I, Horikoshi N, Usheva A. Another function for the mitochondrial ribosomal RNA: Protein folding. Biochemistry 2001; 40:11559-11564.
5. Van Blerkom J, Davis P, Alexander S. Differential mitochondrial distribution in human pronuclear embryos leads to disproportionate inheritance between blastomeres: Relationship to microtubular organization, ATP content and competence. Hum Reprod 2000; 15:2621-2633.
6. Boiani M, Gambles V, Schöler H. ATP levels in cloned mouse embryos. Cytogenet Genome Res 105:270-278.
7. Westermann B. Merging mitochondria matters: Cellular role and molecular machinery of mitochondrial fusion. EMBO Rep 2002; 3:527-531.
8. Chen H, Chan DC. Emerging functions of mammalian mitochondrial fusion and fission. Hum Mol Genet 2005; 14:R283-289.
9. Chen XJ, Butow RA. The organization and inheritance of the mitochondrial genome. Nat Rev Genet 2005; 6:815-825.
10. Anderson S, Bankier AT, Barrell BG et al. Sequence and organization of the human mitochondrial genome. Nature 1981; 290:457-465.
11. GOBASE. The Organelle Genome Database. 2005, (http://www.bch.umontreal.ca/ogmp/projects/other/mt_list.html).
12. Hiendleder S, Wolf E. The mitochondrial genome in embryo technologies. Reprod Domest Anim 2003; 38:290-304.
13. Hiendleder S, Kaupe B, Wassmuth R et al. Molecular analysis of wild and domestic sheep questions current nomenclature and provides evidence for domestication from two different subspecies. Proc Biol Sci 2002; 269:893-904.
14. Lott MT, Brandon M, Brown MD et al. MITOMAP: A human mitochondrial genome database. 2003, (http://www.mitomap.org).
15. Cotter D, Guda P, Fahy E et al. MitoProteome: Mitochondrial protein sequence database and annotation system. Nucleic Acids Res 2004; 32(Database issue):D463-467.
16. Iborra FJ, Kimura H, Cook PR. The functional organization of mitochondrial genomes in human cells. BMC Biol 2004; 2:9.
17. Hutchison IIIrd CA, Newbold JE, Potter SS et al. Maternal inheritance of mammalian mitochondrial DNA. Nature 1974; 251:536-538.
18. Gyllensten U, Wharton D, Wilson AC. Maternal inheritance of mitochondrial DNA during backcrossing of two species of mice. J Hered 1985; 76:321-324.
19. May-Panloup P, Chretien MF, Savagner F et al. Increased sperm mitochondrial DNA content in male infertility. Hum Reprod 2003; 18:550-556.

20. May-Panloup P, Chretien MF, Jacques C et al. Low oocyte mitochondrial DNA content in ovarian insufficiency. Hum Reprod 2005; 20:593-597.
21. Tamassia M, Nuttinck F, May-Panloup P et al. In vitro embryo production efficiency in cattle and its association with oocyte adenosine triphosphate content, quantity of mitochondrial DNA, and mitochondrial DNA haplogroup. Biol Reprod 2004; 71:697-704.
22. Barritt JA, Kokot M, Cohen J et al. Quantification of human ooplasmic mitochondria. Reprod Biomed Online 2002; 4:243-247.
23. Gyllensten U, Wharton D, Josefsson A et al. Paternal inheritance of mitochondrial DNA in mice. Nature 1991; 352:255-7.
24. Kaneda H, Hajashi JI, Takahama S et al. Elimination of paternal mitochondrial DNA in intraspecific crosses during early mouse embryogenesis. Proc Natl Acad Sci USA 1995; 92:4542-4546.
25. Sutovsky P, Moreno RD, Ramalho-Santos J et al. Ubiquitin tag for sperm mitochondria. Nature 1999; 402:371-372.
26. Sutovsky P, Van Leyen K, McCauley T et al. Degradation of paternal mitochondria after fertilization: Implications for heteroplasmy, assisted reproductive technologies and mtDNA inheritance. Reprod Biomed Online 2004; 8:24-33.
27. Schwartz M, Vissing J. Paternal inheritance of mitochondrial DNA. New Engl J Med 2002; 347:576-580.
28. Kraytsberg Y, Schwartz M, Brown TA et al. Recombination of human mitochondrial DNA. Science 2004; 304:981.
29. Zsurka G, Kraytsberg Y, Kudina T et al. Recombination of mitochondrial DNA in skeletal muscle of individuals with multiple mitochondrial DNA heteroplasmy. Nat Genet 2005; 37:873-877.
30. Evans MJ, Gurer C, Loike JD et al. Mitochondrial DNA genotypes in nuclear transfer-derived cloned sheep. Nat Genet 1999; 23:90-93.
31. Loi P, Ptak G, Barboni B et al. Genetic rescue of an endangered mammal by cross-species nuclear transfer using post-mortem somatic cells. Nat Biotechnol 2001; 19:962-964.
32. Steinborn R, Schinogl P, Zakhartchenko V et al. Mitochondrial DNA heteroplasmy in cloned cattle produced by fetal and adult cell cloning. Nat Genet 2000; 25:255-257.
33. Do JT, Lee JW, Lee BY et al. Fate of donor mitochondrial DNA in cloned bovine embryos produced by microinjection of cumulus cells. Biol Reprod 2002; 67:555-560.
34. Steinborn R, Schinogl P, Wells DN et al. Coexistence of Bos taurus and B. indicus mitochondrial DNAs in nuclear transfer-derived somatic cattle clones. Genetics 2002; 162:823-829.
35. Hiendleder S, Zakhartchenko V, Wenigerkind H et al. Heteroplasmy in bovine fetuses produced by intra- and inter-subspecific somatic cell nuclear transfer: Neutral segregation of nuclear donor mitochondrial DNA in various tissues and evidence for recipient cow mitochondria in fetal blood. Biol Reprod 2003; 68:159-166.
36. Takeda K, Akagi S, Kaneyama K et al. Proliferation of donor mitochondrial DNA in nuclear transfer calves (Bos taurus) derived from cumulus cells. Mol Reprod Dev 2003; 64:429-437.
37. Han ZM, Chen DY, Li JS et al. Mitochondrial DNA heteroplasmy in calves cloned by using adult somatic cell. Mol Reprod Dev 2004; 67:207-214.
38. Theoret CL, Dore M, Mulon PY et al. Short- and long-term skin graft survival in cattle clones with different mitochondrial haplotypes. Theriogenology 2006; 65:1465-1479.
39. Jiang Y, Liu SZ, Zhang YL. The fate of mitochondria in Ibex-hirus reconstructed early embryos. Acta Biochim Biophys Sin 2004; 36:371-374.
40. St John JC, Moffatt O, D'Souza N. Aberrant heteroplasmic transmission of mtDNA in cloned pigs arising from double nuclear transfer. Mol Reprod Dev 2005; 72:450-460.
41. Takeda K, Tasai M, Iwamoto M et al. Transmission of mitochondrial DNA in pigs and progeny derived from nuclear transfer of Meishan pig fibroblast cells. Mol Reprod Dev 2006; 73:306-312.
42. Inoue K, Ogonuki N, Yamamoto Y et al. Tissue-specific distribution of donor mitochondrial DNA in cloned mice produced by somatic cell nuclear transfer. Genesis 2004; 39:79-83.
43. Polejaeva IA, Chen SH, Vaught TD et al. Cloned pigs produced by nuclear transfer from adult somatic cells. Nature 2000; 407:86-90.
44. Battersby BJ, Loredo-Osti JC, Shoubridge EA. Nuclear genetic control of mitochondrial DNA segregation. Nat Genet 2003; 33:183-186.
45. Loftus RT, MacHugh DE, Bradley DG et al. Evidence for two independent domestications of cattle. Proc Natl Acad Sci USA 1994; 91:2757-2761.
46. Luikart G, Gielly L, Excoffier L et al. Multiple maternal origins and weak phylogeographic structure in domestic goats. Proc Natl Acad Sci USA 2001; 98:5927-5932.
47. Giuffra E, Kijas J, Amarger V et al. The origin o fthe domestic pig: independent domestication and subsequent introgression. Genetics 2000; 154:1785-1791.

48. Lanza RP, Cibelli JB, Diaz F et al. Cloning of an endangered species (Bos gaurus) using interspecies nuclear transfer. Cloning 2000; 2:79-90.
49. Chen DY, Wen DC, Zhang YP et al. Interspecies implantation and mitochondria fate of panda-rabbit cloned embryos. Biol Reprod 2002; 67:637-642.
50. Yang CX, Han ZM, Wen DC et al. In vitro development and mitochondrial fate of macaca-rabbit cloned embryos. Mol Reprod Dev 2003; 65:396-401.
51. Yang CX, Kou ZH, Wang K et al. Quantitative analysis of mitochondrial DNAs in macaque embryos reprogrammed by rabbit oocytes. Reproduction 2004; 127:201-205.
52. Chen Y, He ZX, Liu A et al. Embryonic stem cells generated by nuclear transfer of human somatic nuclei into rabbit oocytes. Cell Res 2003; 13:251-263.
53. Jiang Y, Chen T, Wang K et al. Different fates of donor mitochondrial DNA in bovine-rabbit and cloned bovine-rabbit reconstructed embryos during preimplantation development. Front Biosci 2006; 11:1425-1432.
54. Jiang Y, Chen T, Nan CL et al. In vitro culture and mtDNA fate of ibex-rabbit nuclear transfer embryos. Zygote 2005; 13:233-240.
55. Chang KH, Lim JM, Kang SK et al. Blastocyst formation, karyotype, and mitochondrial DNA of interspecies embryos derived from nuclear transfer of human cord fibroblasts into enucleated bovine oocytes. Fertil Steril 2003; 80:1380-1387.
56. Li Y, Dai Y, Du W et al. Cloned endangered species takin (Budorcas taxicolor) by inter-species nuclear transfer and comparison of the blastocyst development with yak (Bos grunniens) and bovine. Mol Reprod Dev 2006; 73:189-195.
57. Sansinena M, Lynn J, Denniston R et al. Ooplasmic transfer after interspecies nuclear transfer: Presence of foreign mitochondria, pattern of migration and effect on embryo development. Reprod Fert Dev 2005; 17:182.
58. Richly E, Leister D. NUMTs in sequenced eukaryotic genomes. Mol Biol Evol 2004; 21:1081-108.
59. Parfait B, Rustin P, Munnich A et al. Coamplification of nuclear pseudogenes and assessment of heteroplasmy of mitochondrial DNA mutations. Biochem Biophys Res Commun 1998; 247:57-59.
60. Sorenson MD, Fleischer RC. Multiple independent transpositions of mitochondrial DNA control region sequences to the nucleus. Proc Natl Acad Sci USA 1996; 93:15239-15243.
61. Hiendleder S, Lewalski H, Wassmuth R et al. The complete mitochondrial DNA sequence of the domestic sheep (Ovis aries) and comparison with the other major ovine haplotype. J Mol Evol 1998; 47:441-448.
62. Tully LA, Parsons TJ, Steighner RJ et al. A sensitive denaturing gradient-Gel electrophoresis assay reveals a high frequency of heteroplasmy in hypervariable region 1 of the human mtDNA control region. Am J Hum Genet 2000; 67:432-443.
63. Hiendleder S, Bebbere D, Zakhartchenko V et al. Maternal-fetal transplacental leakage of mitochondrial DNA in bovine nuclear transfer pregnancies: Potential implications for offspring and recipients. Cloning Stem Cells 2004; 6:150-156.
64. Jansen RP. Germline passage of mitochondria: Quantitative considerations and possible embryological sequelae. Hum Reprod 2000; 15(Suppl 2):112-128.
65. Takeda K, Tasai M, Iwamoto M et al. Microinjection of cytoplasm or mitochondria derived from somatic cells affects parthenogenetic development of murine oocytes. Biol Reprod 2005; 72:1397-1404.
66. Tecirlioglu RT, Cooney MA, Lewis IM et al. Comparison of two approaches to nuclear transfer in the bovine: Hand-made cloning with modifications and the conventional nuclear transfer technique. Reprod Fertil Dev 2005; 17:573-585.

CHAPTER 9

Activation of Fertilized and Nuclear Transfer Eggs

Christopher Malcuit and Rafael A. Fissore*

Abstract

In all animal species, initiation of embryonic development occurs shortly after the joining together of the gametes from each of the sexes. The first of these steps, referred to as "egg activation", is a series of molecular events that results in the syngamy of the two haploid genomes and the beginning of cellular divisions for the new diploid embryo. For many years it has been known that the incoming sperm drives this process, as an unfertilized egg will remain dormant until it can no longer sustain normal metabolic processes. Until recently, it was also believed that the sperm was the only cell capable of creating a viable embryo and offspring. Recent advances in cell biology have allowed researchers to not only understand the molecular mechanisms of egg activation, but to exploit the use of pharmacological agents to bypass sperm-induced egg activation for the creation of animals by somatic cell nuclear transfer. This chapter will focus on the molecular events of egg activation in mammals as they take place during fertilization, and will discuss how these mechanisms are successfully bypassed in processes such as somatic cell nuclear transfer.

Introduction

In mammals, ovulation occurs once fully grown oocytes have progressed to a dormant state in the metaphase stage of the second meiotic division (MII). The process leading up to this arrest, termed oocyte maturation, renders the oocyte fully competent to respond to the forthcoming signaling events that take place during fertilization. Part of this preparation period entails the recruitment and translation of stored maternal mRNAs that encode for effectors of the pathways involved in cell cycle progression and calcium (Ca^{2+}) release, and a structural reorganization of key calcium-sensitive elements and organelles.[1-5] The culmination of these events results in a perfectly harmonized system that is poised to respond in the most sensitive and efficient manner to the incoming sperm.

Exit from the MII arrest and meiotic resumption, referred to as "egg activation", is made possible by the fertilizing sperm, which evokes in the egg an increase in the concentration of intracellular free calcium ions ($[Ca^{2+}]_i$). This rise in $[Ca^{2+}]_i$ is necessary and sufficient for the completion of all the events of egg activation,[6] including exocytosis of the cortical granule material to block polyspermy,[7] resumption of the meiotic cell cycle through ubiquitin-proteasome-dependent destruction of cyclin B,[8] pronuclear formation, recruitment of maternal mRNAs, and initiation of mitotic divisions that unveil the complete developmental program.[6,9]

*Corresponding Author: Rafael A. Fissore—Department of Veterinary and Animal Sciences, Paige Laboratory, University of Massachusetts, Amherst, Massachusetts 01003; U.S.A. Email: rfissore@vasci.umass.edu

Somatic Cell Nuclear Transfer, edited by Peter Sutovsky. ©2007 Landes Bioscience and Springer Science+Business Media.

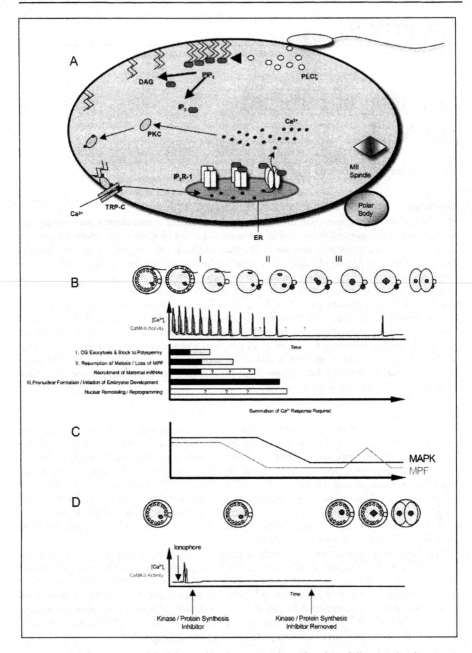

Figure 1. A) Schematic model of the molecular events that take place following fertilization in mammals. A soluble protein, presumably phospholipase Cζ (PLCζ) diffuses into the ooplasm after fusion of the gametes. PLCζ then catalyzes the hydrolysis of PIP$_2$ into the two signaling molecules di-acyl glycerol (DAG) and inositol trisphosphate (IP$_3$). IP$_3$ binds and un-gates its receptor, the IP$_3$R-1, located on the lumen of the endoplasmic reticulum (ER), thereby causing Ca^{2+} efflux into the cytosol. Figure legend continued on next page.

Figure 1 Continued. Cytosolic Ca^{2+} activates a protein kinase C (PKC), thereby targeting it to the plasma membrane where it acts in further Ca^{2+} influx into the egg through further action by DAG, possibly through transient receptor potential channels (TRP-C). Ca^{2+} influx is regulated by store-operated Ca^{2+} entry (SOCE) that refills the ER, and in the presence of persistent IP_3 production (as is the case during fertilization) results in an oscillatory system. B) Representative trace profile of the pattern of $[Ca^{2+}]_i$ oscillations (black) seen after fertilization and egg activation procedures that employ strontium chloride or cytosolic components of sperm. $[Ca^{2+}]_i$ oscillations persist in mammals for several hours, and are accompanied by repetitive rises in the activity levels of Ca^{2+}-calmodulin protein kinase type II (CaMKII, blue trace), which is responsible for transducing the Ca^{2+} signal into events of egg activation, represented by the bar graph. The bar graph is a representation of the persistence of $[Ca^{2+}]_i$ oscillations required to achieve particular levels of egg activation events, as outlined by Ducibella (et al, 2002) and Ozil (et al, 2005). The black solid portions of the horizontal bars denote the ~ number of $[Ca^{2+}]_i$ rises required to initiate an activation event, whereas the shaded areas within the bar ~ denotes the number of $[Ca^{2+}]_i$ rises needed to complete each event. Egg activation events in which a direct link to the Ca^{2+} activation stimulus needs further demonstration are marked with a question mark (?) inside the corresponding bar. C) An approximate trace of the activity levels of maturation promoting factor (MPF) and mitogen activating protein kinase (MAPK) levels during egg activation and entry into the first mitotic cell cycle. Time course coincides roughly with the Ca^{2+} profile depicted in both B and D (above and below). Each Ca^{2+} transient has been reported to result in cumulative destruction of cyclin B, thereby resulting in the drop in the levels of active MPF. Shortly after levels of MPF have subsided, MAPK levels begin to decrease, and reach basal levels at around the time of pronuclear formation as well as the termination of $[Ca^{2+}]_i$ oscillations. D) Depiction of the traces of $[Ca^{2+}]_i$ (black) and CaMKII (blue) levels during a technique used to activate somatic cell nuclear transfer embryos. Note that the use of protein kinase or synthesis inhibitors makes further oscillations in the levels of $[Ca^{2+}]_i$ obsolete, at least for progression of the cell cycle. However, it appears clear that several hours of complex cellular signaling events are bypassed in this type of procedure, which was the focus of this chapter.

Although the precise mechanism by which the sperm initiates the Ca^{2+} release that is responsible for triggering embryonic developmental is not yet fully clear, it has been established in all species studied to date that it involves the activation of the phosphoinositide (PI) pathway (Fig. 1A).[10,11] Activation of the PI pathway in eggs results in the production of inositol 1,4,5-trisphosphate (IP_3) and 1,2-diacylglycerol (DAG) via the hydrolysis of phosphatidyl 4,5-bisphosphate (PIP_2) by a phosphoinositide-specific phospholipase C (PLC) isoform.[12-14] Both products of the PI pathway are involved in shaping $[Ca^{2+}]_i$ responses. Increase in the intracellular concentrations of IP_3 is responsible for mediating Ca^{2+} release by binding and gating its receptor, the type I IP_3 receptor (IP_3R-1),[15] a tetrameric ligand-gated Ca^{2+} channel located on the endoplasmic reticulum (ER) membrane, the main Ca^{2+} store of the cell.[16-17] Production of DAG, either directly[18] or indirectly via activation of protein kinase C (PKC), may be involved in the regulation of Ca^{2+} influx.[19] Notably, in spite of the universal requirement for a $[Ca^{2+}]_i$ increase in egg activation, the Ca^{2+} release evoked by fertilization vary widely among species. For instance, in some lower vertebrates and marine animals Ca^{2+} release takes the form of a single, rather long (10 minutes) transient, whereas mammalian eggs show persistent and repetitive changes in $[Ca^{2+}]_i$ (reviewed in ref. 20) beginning shortly after fertilization[21] and lasting for several hours. Interestingly, in mouse eggs, sperm-initiated $[Ca^{2+}]_i$ responses cease in concurrence with pronuclear formation,[22-24] while in other mammalian species oscillations persist throughout the first cell cycle.[25,26] Therefore, the differences in $[Ca^{2+}]_i$ responses among phyla and even within classes suggest, at the very least, evolutionary divergence of the mechanism(s) leading to activation of the PI pathway.

Mechanisms of Sperm-Induced Ca^{2+} Release

There has been much debate and speculation as to the mechanism(s) that triggers activation of the PI pathway and $[Ca^{2+}]_i$ oscillations during mammalian fertilization, and this has led to the proposal of several theories to explain this phenomenon. One early hypothesis, referred to as the "conduit hypothesis", proposed that sperm fusion allows Ca^{2+} to passively enter the egg. However, the findings that the initial fertilization-induced Ca^{2+} responses proceed unaltered in the absence of extracellular Ca^{2+} ($[Ca^{2+}]_e$),[27] clearly demonstrated that $[Ca^{2+}]_e$ is not necessary for the initiation of $[Ca^{2+}]_i$ oscillations. The "receptor hypothesis" suggests that upon sperm-egg membrane contact, receptor-ligand interactions on the surface of the gametes relay intracellular signaling events that initiate Ca^{2+} release in the egg. One of the signaling cascades thought to be engaged by the interaction of gametes is that mediated by protein tyrosine kinases (PTKs). Specifically, the Src-family of PTKs (SFKs) may activate PLCγs,[28-35] thereby triggering Ca^{2+} release through the production of IP_3. In accordance with this notion, PTK and PLCγ activity are up-regulated shortly following fertilization in echinoderm and amphibian eggs,[13,36] and fertilization-triggered Ca^{2+} signaling in egg extracts was reconstituted by adding "activated" membrane raft fractions,[33] which suggests the presence of a receptor-mediated SFK/PLCγ activation model in Xenopus fertilization. Furthermore, inhibition of PLCγ activation by a dominant negative approach using over expression of PLCγ SH2 domains prevented the sperm-induced $[Ca^{2+}]_i$ rise in sea urchin[28] and starfish.[37] Nonetheless, extensive pharmacological studies[38] along with dominant negative approaches[39] and injection of recombinant PLCγ[40] failed to show any involvement of this pathway in evoking $[Ca^{2+}]_i$ responses in rodent eggs. This is in spite of the fact that PLCγ isoforms are expressed in mouse eggs,[41,42] and that stimulation of its activity in these cells by exogenous expression of tr-kit, a sperm tyrosine kinase which activates Fyn, a SFK present in mouse and rat eggs,[43] was able to induce exit from meiosis and pronuclear formation.[44] Therefore, while we can discount a role for SFKs on the initiation of $[Ca^{2+}]_i$ oscillations in rodent fertilization, possible downstream effects of this signaling cascade on other events of egg activation need further investigation,[45] particularly in nonrodent mammals. Lastly, extensive studies, which initially relied on the injection of activators and inhibitors of α subunits of G_q proteins, implicated the activation of PLCβ isoforms in the initiation of Ca^{2+} release in mammalian fertilization.[46-48] However, the subsequent findings demonstrating that inhibition of $Gα_q$ subunits by injection of a function-blocking antibody was without effect on fertilization-induced $[Ca^{2+}]_i$ oscillations,[49] in conjunction with the apparent normal fertility in most strain of mice lacking one of the PLCβ isoforms,[50] suggest that the contribution of this pathway to the initiation of oscillations in mammalian fertilization is negligible.

Given the inability of the aforementioned experimental paradigms to recapitulate the initiation of $[Ca^{2+}]_i$ oscillations in mammals, consensus began to coalesce on the need for a novel mechanism to explain the initiation of oscillations in these species. The hypothesis that emerged, "the fusion hypothesis", proposed that upon gamete fusion the sperm delivers a factor, commonly referred to as the sperm factor (SF), into the ooplasm capable of activating the PI pathway and oscillations.[51] The initial, and sole, experimental support for this hypothesis was the demonstration that injection of sperm extracts into mammalian eggs was able to replicate the pattern of oscillations initiated by the sperm.[52-55] Curiously, the first demonstration of this mechanism was obtained in sea urchin eggs,[56] a species in which, paradoxically, the hypothesis under discussion may not account for the mechanism of fertilization (see above). Concomitantly, the advent of intracytoplasmic sperm injection,[57] a technique by which an intact sperm delivered into the ooplasm is capable of initiating fertilization-like $[Ca^{2+}]_i$ responses[58] and embryo development to term, consolidated the concept that a sperm-cytosolic molecule was responsible for initiating oscillations in mammalian eggs.

Molecular Identity of the Sperm's Activating Factor

Despite the fact that the identity of the molecule(s) responsible for the SF activity has yet to be fully ascribed, glimpses of the properties of the putative molecule(s) have emerged from fertilization and biochemical studies.[14,54,59-62] First, while early studies assumed that all SF activity was rapidly released into the ooplasm,[52,53] subsequent studies revealed that the Ca^{2+}-inducing activity was also present in detergent-resistant sperm domains, most likely associated with the sperm perinuclear theca.[59,60,63-66] Congruent with the concept of SF distribution to several sperm compartments was the demonstration that complete release of SF activity into the ooplasm demanded ~2 hours.[66] Second, consistent with its perinuclear localization in the sperm, in vitro fertilization and ICSI studies showed that sperm's Ca^{2+}-releasing activity could be recovered after fertilization, as it associated with the pronuclei of the developing zygotes.[66-68] Third, in vitro PLC assays using sperm extracts showed that these extracts possessed high PLC activity, nearly twice as high as the activity present in other tissues known to express several PLC isoforms.[14,61] Importantly, the PLC activity of sperm extracts is prominent even in the presence of low levels of $[Ca^{2+}]_e$, ~0.1 μM, which is very relevant given that this molecule is expected to initiate oscillations in mammalian eggs and that this is the basal $[Ca^{2+}]_i$ level in eggs prior to fertilization. Therefore, since several PLC isoforms are expressed in mammalian sperm,[50,62,69,70] these enzymes surfaced as logical candidates to be a component of the SF. Importantly, injection of recombinant proteins representing most of the isoforms expressed in sperm failed to initiate oscillations, or it did so at non physiological concentrations.[40,61] Furthermore, chromatographic fractionation of sperm extracts revealed that none of the known PLCs were present in the fractions with $[Ca^{2+}]_i$ oscillation-inducing activity.[62,70] Hence, if a sperm PLC were to be the SF, it had to be a novel PLC. Towards this end, a novel sperm-specific PLC, PLCζ,[71] was screened out of mouse expressed sequence tag libraries. Initial studies revealed that PLCζ exhibits $[Ca^{2+}]_i$ oscillation-inducing activity ascribed, thus far, only to the sperm or SF. Hence, PLCζ may be the long sought-after oscillogenic component of the SF.

Since its discovery from mouse testis,[71] PLCζ has also been identified and cloned in humans and nonhuman primates,[72] and highly homologous sequences have been reported for several other species including bovine and porcine (Genbank accession numbers GI:55669158 and GI:32400660, respectively). PLCζ is currently the most elementary of PLC isoforms identified. In concurrence with the modular organization of other PLCs,[73] PLCζ consists of two Ca^{2+}-binding EF hands, X and Y catalytic domains, and the Ca^{2+}-dependent phospholipid-binding C2 domain.[74] Notably, PLCζ lacks the typical pleckstrin homology (PH) domain, which has been found in all previously identified PLC isoforms.[75] In support of its purported role as the SF, injection of recombinant PLCζ[76,77] or PLCζ cRNA has been shown to evoke fertilization-like- oscillations in mouse,[71,72,76-78] human,[79] and bovine eggs.[80] In addition, in vitro PLC assays using recombinant PLCζ revealed high enzymatic activity at basal (~ 0.1 μM) $[Ca^{2+}]_e$ concentrations[76] and the regulatory effect of the EF hand and C2 domains on the enhanced Ca^{2+} sensitivity of the enzyme and its oscillatory activity in mouse eggs.[81,82] Moreover, PLCζ cRNA-induced oscillations cease at the time roughly corresponding to pronuclear formation, which is comparable to what is observed after natural fertilization in mouse[78,83] and, like the sperm-induced oocyte activation, zygotes activated by injection of PLCζ cRNA showed high rates of in vitro development to the blastocyst stage.[71,72] Lastly, a recent report in mouse sperm has tentatively localized PLCζ to the post-acrosomal region of mouse sperm,[77] which is the first area thought to come in contact with the ooplasm after the fusion of gametes.[65]

How Is the Ca^{2+} Signal Translated into Events of Egg Activation?

In spite of the realization that the activation of embryonic development in mammals relies on the initiation of $[Ca^{2+}]_i$ oscillations, there is remarkably little known as to what are the "molecular effectors" in the egg that translate the transient $[Ca^{2+}]_i$ elevations into cellular events. It is well established that unlike echinoderm and amphibian fertilization, initiation of multiple rises is required to promote complete exit from meiosis in mammals (for review see ref. 20). Subsequent studies, first in Xenopus eggs, have lent support to the notion that $[Ca^{2+}]_i$ elevations mediate meiotic resumption by activating the type II Ca^{2+}/calmodulin-dependent protein kinase (CaMKII),[84] which in turn is responsible for initiating a cascade that leads to activation of the anaphase promoting complex/cyclosome (APC/C).[85,86] The APC, an E3 ubiquitin ligase complex, targets cell cycle proteins such as cyclin B for proteasomal degradation following their poly-ubiquitination.[87-91] The loss of cylin B, the regulatory subunit of MPF[92] and limiting factor for the maintenance of the metaphase state, results in the exit from M-phase and thus the resumption of meiosis. Importantly, in mammals, each $[Ca^{2+}]_i$ rise results in a concomitant activation of CaMKII,[93,94] and this is envisioned to be responsible for the steady increase in the rate of cyclin degradation through maintenance of APC/C activity.[91,95] Interestingly, this Ca^{2+}-mediated cell-cycle transition is thought to be unique to eggs,[91] and requires stable levels of cytostatic factor (CSF),[8] which is itself unique to eggs.[96]

If the only function of $[Ca^{2+}]_i$ oscillations were to promote exit of meiosis, then elevations of $[Ca^{2+}]_i$ according to disparate, but not excessive protocols, should lead to equal rates of activation and development. Nonetheless, elegant studies by Ozil and coworkers[97-99] have demonstrated that this is not the case. In their studies, parthenogenetic activation protocols that adhered more closely to the fertilization-like patterns of oscillations resulted in higher rates of implantation and development. In these studies, the amplitude, frequency, and number of $[Ca^{2+}]_i$ rises were modulated by electroporation of $[Ca^{2+}]_e$ into the ooplasm of rabbit and mouse eggs. Each parameter tested had a marked effect on the rates of egg activation as well as on post-implantation development. In line with the developmental benefits of Ca^{2+}, more recent studies revealed that different events of egg activation have dissimilar requirements regarding their initiation and completion, with the completion of each event needing higher number of rises (Fig. 1B, C).[9,99] Especially relevant to embryo development, is the possible impact of the $[Ca^{2+}]_i$ activating stimulus on the recruitment of specific maternal mRNAs,[9,99] the recruitment and translations of which may be critical for the activation of the zygotic genome.[100] For instance, it was shown that "recruitment of maternal mRNAs", as evidenced by the detection of new protein synthesis, was initiated by the administration of 8 $[Ca^{2+}]_e$ pulses but only became fully apparent, or "fertilization-like", in zygotes receiving a total of 24 $[Ca^{2+}]_e$ pulses.[9] While these results insinuate a relationship between $[Ca^{2+}]_i$ oscillations and protein synthesis, it is important to discriminate whether the observed changes in protein profiles were not simply due to progression into zygotic interphase, which seemed to exhibit similar requirements for $[Ca^{2+}]_e$ pulses.[9] Nonetheless, the approach used in this study, followed by identification of the proteins regulated by Ca^{2+} pulses, could prove invaluable in elucidating yet uncharacterized signaling pathways during egg activation.

How then does the egg transduce these repetitive $[Ca^{2+}]_i$ rises into an effector? It has recently been shown that just as concentrations of $[Ca^{2+}]_i$ oscillate in response to IP_3 production, so too does the activity of CaMKII.[93,94,101] For instance, the activity of CaMKII not only increased in response to augmenting levels of $[Ca^{2+}]_i$, but oscillated in close synchrony with each $[Ca^{2+}]_i$ transient after fertilization, as determined in single eggs by simultaneous monitoring of $[Ca^{2+}]_i$ levels and kinase activity (Fig. 1D).[93,94] Hence, given that each $[Ca^{2+}]_i$ transient during fertilization accounts for a single event during which CaMKII acts upon its targets, it can be speculated that the compounding effect of each $[Ca^{2+}]_i$ rise is required for brief periods of CaMKII activity,[102] the culmination of which results in successful egg activation. In light of a recent report demonstrating that constitutively active CaMKII is sufficient for the resumption of meiosis in mouse eggs,[103] CaMKII appears to be the single most important messenger identified thus far in the pathway of Ca^{2+}-generated cell-cycle progression during egg activation. Importantly, a recent report in hippocampal dendrites has demonstrated that CaMKII directly phosphorylates the cytoplasmic polyadenylation element binding protein (CPEB) that is directly responsible for the polyadenylation, recruitment, and translation of mRNAs.[104] A similar mechanism therefore, could potentially function in eggs. This would provide a direct link between the results mentioned above in which repetitive $[Ca^{2+}]_i$ pulses modulated protein expression patterns from stored maternal mRNAs.[9,99] It is worth noting that other kinases such as PKC,[19] or proteins such as actin[105] and calreticulin,[106] may also serve as effectors of $[Ca^{2+}]_i$ rises during egg activation, although additional research is required to realize the extent of their participation.

Activation of Somatic Cell Nuclear Transfer Embryos

The increased understanding of the molecular mechanisms underlying egg activation set the stage for the development of effective methods of egg parthenogenetic activation. The application of these activation procedures has been indispensable for the generation of embryos and offspring by somatic cell nuclear transfer (SCNT). To date, multiple animal species including sheep,[107] cattle,[108] goat,[109] mouse,[110] pig,[111] rabbit,[112] cat,[113] mule,[114] horse,[115] and rat[116] have been successfully cloned using somatic cells and activation stimuli independent from the sperm.

There are several approaches to exogenous egg activation in mammalian SCNT, all of which have the same goal: exit from meiosis and initiation of embryonic development by inactivation of MPF and MAPK activities. As mentioned above, natural suppression of these kinases, specifically MPF, is achieved through repetitive Ca^{2+} signaling. Without the sperm's contribution, exogenous methods to reduce these kinases must be employed. One approach is to evoke $[Ca^{2+}]_i$ oscillations in eggs comparable to those induced by the sperm. In this regard, the most effective method has been the application of the compound $SrCl_2$ which, when used in place of extracellular Ca^{2+}, simulates $[Ca^{2+}]_i$ oscillations through IP_3R-1, causing egg activation in a manner very similar to fertilization.[117] This method has been the one of choice for activation of mouse nuclear transfer embryos.[118] $SrCl_2$ is thought to act by sensitizing IP_3R-1 to basal-intracellular concentrations of IP_3 causing, in-turn, the gating of IP_3R-1 and Ca^{2+} release.[119] However, $SrCl_2$ does not act in a manner indistinguishable from the sperm, as the duration of the rises in $[Ca^{2+}/Sr^{2+}]_i$ levels observed during $SrCl_2$-induced oscillations is considerably longer than those during fertilization and the IP_3R-1 content is not reduced by this treatment,[120] demonstrating that $SrCl_2$ bypasses activation of the PI pathway. The developmental impact of this, especially the presumed lack of activation of PKC,[19] is presently not known. Nonetheless, even though $SrCl_2$ requires cotreatment with a microfilament inhibitor such as a cytochalasin,[118] it is the preferred method of activation for mouse cloning, as it alleviates the need for supplemental compounds that may have more widespread and possibly deleterious effects on embryonic development. Unfortunately, $SrCl_2$ does not elicit $[Ca^{2+}/Sr^{2+}]_i$ oscillations in bovine eggs (our unpublished results) and, most likely, in eggs of other large domestic species, precluding its use for cloning procedures in these species. Alternatively, in these species, $[Ca^{2+}]_i$ oscillations can be initiated by other methods such as the injection of sperm extracts or IP_3R-1 agonists.[121] However, while injection of sperm extracts consistently induced oscillations and supported full term development of SCNT reconstructed zygotes, it required microinjection and the initiated $[Ca^{2+}]_i$ oscillations were short-lived.[121]

Generation of $[Ca^{2+}]_i$ oscillations has also been attempted by the administration of electrical DC pulses. In the presence of exogenous Ca^{2+}, electroporation induces Ca^{2+} influx thereby transiently elevating $[Ca^{2+}]_i$ in the ooplasm,[47] as the activation of Ca^{2+} pumps in the endoplasmic reticulum (ER), the Ca^{2+} store of cells, and in the plasma membrane return $[Ca^{2+}]_i$ levels to basal concentrations. The delivery of DC pulses allows exquisite regulation of $[Ca^{2+}]_i$ levels in eggs[97] and has resulted in very effective parthenogenetic development of parthenogenetic embryos generated by this approach.[98,99] Furthermore, this method of activation has been shown to be successful in the cloning of the first sheep,[122] pigs[111] and goats,[109] even though these studies administered a single DC pulse. However, in cattle, whose eggs seem to require multiple $[Ca^{2+}]_i$ rises to become activated, electrical stimulation has not gained widespread use as the method of choice for activation of SCNT embryos, as the delivery of multiple pulses with the technology presently available is cumbersome. Thus, another strategy was required for downregulating MPF and MAPK activities. The use of the broad-spectrum ser/thr kinase inhibitor 6-dimethylaminopurine (6-DMAP) had been described to induce high rates of bovine egg activation and in vitro embryo development.[123] Consequently, this inhibitor was used to clone the first calves.[108] 6-DMAP acts by inhibiting a critical phosphorylation event on cdc25, which is required for its activation;[124] cdc25 is the phosphatase responsible for activating MPF from pre-MPF.[125] Cdc25 acts by dephosphorylating Tyr15 and Thr14 (the inhibitory phosphates) of pre-MPF. By inhibiting cdc25, pre-MPF fails to become active and thus the metaphase state cannot be maintained. Additionally, 6-DMAP is responsible for dephosphory-lation of ERK-2 (a MAPK), a component of CSF in the MII egg.[126] Through these mecha-nisms, 6-DMAP has proven to be an effective inducer of cell cycle progression in oocytes[127] and embryonic development in SCNT embryos.[108,123,128] However, a disadvantage of using 6-DMAP is its lack of specificity as well as its reported increase of chromosomal abnormalities in embryos produced by this method.[129] Given that in addition to MPF and MAPK many other kinases and phosphatases may be affected by 6-DMAP, a variety of presently unrecog-nized cellular processes may be inhibited/altered in zygotes/embryos activated by this com-pound; whether these additional effects compromise embryo development is not known. How-ever, pharmacological research has also yielded several other small molecules that exhibit higher specificity than 6-DMAP, as they target cyclin-dependent protein kinases (cdks) including cdk1, which is the catalytic member of the MPF complex.[92] Compounds such as roscovitine, olomoucine, and bohemine are purine analogs that specifically compete for ATP binding sites on cdk molecules.[130] However, extensive research on the use of these compounds as an alterna-tive to 6-DMAP in SCNT has not yet been reported in the literature.

Cyclohexamide (CHX), a general protein synthesis inhibitor that acts by stalling mRNA translation on the ribosome, has also been utilized in SCNT, specifically in cattle in which it has been shown to produce live offspring.[131] CHX acts by limiting the available amount of cyclin B in the zygote,[132] the regulatory subunit of MPF, precluding the continuous formation of active MPF, which is required for MII arrest. However, as much as 6-DMAP is nonspecific in its mode of action on kinases and phosphatases, CHX is nonspecific in its action on protein synthesis. Therefore, while zygotes are incubated in the presence of CHX, other protein syn-thesis that may be required for proper embryonic development or nuclear reprogramming may be also abrogated, or their synthesis delayed, possibly disrupting the temporal pattern of gene expression in the early embryo.

It should be noted, however, that the use of agents such as 6-DMAP, roscovitine, or CHX for activation of eggs of large domestic species still requires an initial Ca^{2+} rise. In most com-monly used protocols, this is usually provided by the application of a Ca^{2+} ionophore such as A23187 or ionomycin.[108,123] Ionophores act by making possible the passive diffusion of Ca^{2+} ions across the plasma membrane and ER into the ooplasm. Thus, it appears clear that al-though the need for $[Ca^{2+}]_i$ oscillations can be bypassed with the supplementation of kinase and protein synthesis inhibitors, the universal requirement for Ca^{2+} during egg activation can-not be neglected or excluded. However, whether or not a fertilization-like pattern of oscilla-tions is required to achieve optimal development of SCNT embryos remains to be demon-strated. The recent findings that PLCζ mRNA injection initiates fertilization-like oscillations in bovine eggs[80] should facilitate this type of investigation. Furthermore, ooplasts obtained from zygotes activated by the sperm after IVF were used to clone the first pig, as reported by Polejaeva et al.[111] In their procedure, reconstructed SCNT embryos were activated by an elec-trical pulse and allowed to develop to the pronuclear (1-cell) stage. At such time, the pro-nucleus of the cloned embryo was removed and transplanted into an enucleated 1-cell zygote that had been produced by IVF. In this way, the cloned nucleus was introduced to an embryo that had been activated by the sperm, resulting in the production of live piglets. Nonetheless, whether or not sperm-like $[Ca^{2+}]_i$ oscillations represent the ideal method of activation for SCNT embryos needs additional investigation.

Implications for Exogenous Activation Methods

An additional question that remains to be answered is how the bypassing of physiological $[Ca^{2+}]_i$ oscillations may affect the development of SCNT embryos. Although it has been re-ported that treatment of pig oocytes with the Ca^{2+} ionophore A23817 was sufficient to induce the cortical reaction and block to polyspermy,[133] a thorough analysis of the effects of the various exogenous activation protocols on the parameters of "egg activation" outlined by Schultz and Kopf[6] and carefully tested by Ducibella and colleagues[9] in the mouse has yet to be carried out in other mammals. Likewise, the experimental demonstration that zygotes activated by $[Ca^{2+}]_i$ oscillations show developmental advantages over zygotes generated by other, less physiological, activation protocols is based, mostly, on the in vivo development of diploid parthenogenotes.[99] However, whether reprogramming a somatic nucleus, the task at hand during SCNT, requires $[Ca^{2+}]_i$ oscillations and kinases inactivation profiles similar to those needed to transform meiotic nuclei is not known and deserves careful investigation.

Besides the possible aforementioned impact that the Ca^{2+} activating stimulus may have on zygotic gene expression via recruitment and translation of specific maternal mRNAs, an equally intriguing possibility is that the fertilization-initiated $[Ca^{2+}]_i$ responses are also involved in the nuclear remodeling of the parental genomes, which commences soon after fertilization.[134] One of the most difficult facts to reconcile in somatic cell cloning is the low success of live offspring produced.[135,136] There are certainly multiple factors involved in the reduced efficiency of this procedure, as both the sources of genetic material as well as the methods used to activate development differ dramatically from that seen during fertilization. Undoubtedly genomic imprinting deficiencies caused by nuclear transplantation likely account for a large percentage of the losses seen during cloning. For instance, a report on cloned bovine embryos demonstrated not only differences in certain gene transcripts, but that heat shock protein (Hsp) 70 transcript levels were absent in cloned embryos.[137] Furthermore, this report demonstrated aberrant expression of Mash2 (required for trophoblast formation) and DNA methyltransferase-1 (involved in epigenetic modifications of DNA). However, whether the activation stimulus regulates epigenetic modifications and nuclear remodeling of SCNT embryos or, for that matter, of normally fertilized embryos is not known.

The extensive reconfiguration that gametes undergo soon after fertilization is also required for activation of the zygotic genome (for review see ref. 138). Of the two gametes, the paternal genome undergoes the most rapid and dramatic change as it is subjected to the exchange of protamines for maternal histones, and this occurs in concert with sperm nuclear decondensation.[139,140] Additionally, the paternal genome undergoes near complete DNA demethylation of CpG dinucleotides before reaching the pronuclear stage.[140,141] While the precise nature of the molecular mechanisms that underlie the transformation of the sperm nucleus into a pronucleus is not fully elucidated, given it coincides with presence of $[Ca^{2+}]_i$ oscillations, a role for Ca^{2+}-dependent processes in such events cannot be discounted. In addition, the contribution of other downstream products of the PI pathway also cannot be dismissed, as suggested by the recently demonstrated accumulation of PLCζ to the pronucleus.[142]

Nuclear remodeling and epigenetic modification of the zygotic genome also entails covalent modifications of the nucleosomes' core histones.[134,144] These changes involve phosphorylation, acetylation, and methylation of histones as well as the removal by opposing enzymes of these modifications from the basic amino acids lysine and arginine, which are the target amino acids in these proteins.[145,146] While histone methylases have been identified in somatic cells and eggs, the enzymes that remove these modifications are not presently well known. Importantly, it was reported that mouse oocytes and eggs abundantly express a peptidylarginine demethyliminase-like molecule, ePAD, which functionally demethylates histone by the conversion of methylarginine to citrulline, and the subsequent release of methylimine.[146-148] Relevant to the possible role of Ca^{2+} in nuclear reprogramming is the finding that PADs are Ca^{2+} sensitive enzymes.[148,149] Therefore, it will be important to address the issue of the Ca^{2+}-dependence of ePAD for its functional role in nuclear remodeling and reprogramming of the fertilized zygote and to determine what is the overall effect of the activation procedure on epigenetic modifications and gene expression of SCNT embryos.

Summary

In general, physiological egg activation following fertilization in mammals is driven by the PI system through persistent and long-lasting increases in $[Ca^{2+}]_i$. These repetitive transients of Ca^{2+} release, which activate CaMKII, may be indirectly responsible for the developmentally regulated degradation of cyclin B, the regulatory subunit of MPF, one of the major kinases responsible for the MII arrest. Additionally, Ca^{2+} may signal other downstream effectors through CaMKII, such as the recruitment of stored-maternal mRNAs, and the activation of nuclear remodeling complexes and enzymes controlling the epigenetic fate of the developing embryo, all of which may greatly impact the developmental fate of the zygote.

Through somatic cell nuclear transfer (and to a lesser extent parthenogenesis) it has been demonstrated time after time, and through multiple species, that fertilization is sufficient but not necessary for the production of live offspring. The exogenous egg activation methods reviewed here represent a few ways in which bypassing sperm-induced egg activation results in the production of live offspring. However, the success of SCNT remains very low and embryos exhibit altered methylation and gene expression patterns. Whether the exogenous activation procedure is responsible, at least in part, for these alterations is not known. More importantly, future studies should address whether or not $[Ca^{2+}]_i$ oscillations are at all required for successful SCNT, whether a " distinct activation protocol" is required to reprogram somatic nuclei and, if so, to uncover the downstream intervening Ca^{2+} sensitive molecules that participate in nuclear reprogramming.

References

1. Fujiwara T, Nakada K, Shirakawa H et al. Development of inositol trisphosphate-induced calcium release mechanism during maturation of hamster oocytes. Dev Biol 1993; 156(1):69-79.
2. Shiraishi K, Okada A, Shirakawa H et al. Developmental changes in the distribution of the endoplasmic reticulum and inositol 1,4,5-trisphosphate receptors and the spatial pattern of Ca^{2+} release during maturation of hamster oocytes. Dev Biol 1995; 170(2):594-606.
3. Mehlmann LM, Terasaki M, Jaffe LA et al. Reorganization of the endoplasmic reticulum during meiotic maturation of the mouse oocyte. Dev Biol 1995; 170(2):607-615.
4. Mehlmann LM, Mikoshiba K, Kline D. Redistribution and increase in cortical inositol 1,4,5-trisphosphate receptors after meiotic maturation of the mouse oocyte. Dev Biol 1996; 180(2):489-498.
5. Machaca K. Increased sensitivity and clustering of elementary Ca^{2+} release events during oocyte maturation. Dev Biol 2004; 275(1):170-182.
6. Schultz RM, Kopf GS. Molecular basis of mammalian egg activation. Curr Top Dev Biol 1995; 30:21-62.
7. Cran DG, Moor RM, Irvine RF. Initiation of the cortical reaction in hamster and sheep oocytes in response to inositol trisphosphate. J Cell Sci 1988; 91(Pt 1):139-144.
8. Hyslop LA, Nixon VL, Levasseur M et al. Ca^{2+} promoted cyclin B1 degradation in mouse oocytes requires the establishment of a metaphase arrest. Dev Biol 2004; 269(1):206-219.
9. Ducibella T, Huneau D, Angelichio E et al. Egg-to-embryo transition is driven by differential responses to Ca^{2+} oscillation number. Dev Biol 2002; 250(2):280-291.
10. Turner PR, Sheetz MP, Jaffe LA. Fertilization increases the polyphosphoinositide content of sea urchin eggs. Nature 1984; 310(5976):414-415.
11. Stith BJ, Espinoza R, Roberts D et al. Sperm increase inositol 1,4,5-trisphosphate mass in Xenopus laevis eggs preinjected with calcium buffers or heparin. Dev Biol 1994; 165(1):206-215.
12. Parrington J, Brind S, De Smedt H et al. Expression of inositol 1,4,5-trisphosphate receptors in mouse oocytes and early embryos: The type I isoform is upregulated in oocytes and downregulated after fertilization. Dev Biol 1998; 203(2):451-461.
13. Rongish BJ, Wu W, Kinsey WH. Fertilization-induced activation of phospholipase C in the sea urchin egg. Dev Biol 1999; 215(2):147-154.
14. Rice A, Parrington J, Jones KT et al. Mammalian sperm contain a Ca^{2+}-sensitive phospholipase C activity that can generate InsP3 from PIP2 associated with intracellular organelles. Dev Biol 2000; 228(1):125-135.
15. Miyazaki S, Shirakawa H, Nakada K et al. Essential role of the inositol 1,4,5-trisphosphate receptor/Ca^{2+} release channel in Ca^{2+} waves and Ca^{2+} oscillations at fertilization of mammalian eggs. Dev Biol 1993; 158(1):62-78.

16. Koch GL. The endoplasmic reticulum and calcium storage. Bioessays 1990; 12(11):527-31.
17. Berridge MJ. The endoplasmic reticulum: A multifunctional signaling organelle. Cell Calcium 2002; 32(5-6):235-249.
18. Bazzi MD, Nelsestuen GL. Differences in the effects of phorbol esters and diacylglycerols on protein kinase C. Biochemistry 1989; 28(24):9317-9323.
19. Halet G, Tunwell R, Parkinson SJ et al. Conventional PKCs regulate the temporal pattern of Ca^{2+} oscillations at fertilization in mouse eggs. J Cell Biol 2004; 164(7):1033-1044.
20. Stricker SA. Comparative biology of calcium signaling during fertilization and egg activation in animals. Dev Biol 1999; 211(2):157-176.
21. Lawrence Y, Whitaker M, Swann K. Sperm-egg fusion is the prelude to the initial Ca^{2+} increase at fertilization in the mouse. Development 1997; 124(1):233-241.
22. Jones KT, Carroll J, Merriman JA et al. Repetitive sperm-induced Ca^{2+} transients in mouse oocytes are cell cycle dependent. Development 1995; 121(10):3259-3266.
23. Marangos P, FitzHarris G, Carroll J. Ca^{2+} oscillations at fertilization in mammals are regulated by the formation of pronuclei. Development 2003; 130(7):1461-1472.
24. Jellerette T, Kurokawa M, Lee B et al. Cell cycle-coupled $[Ca^{2+}]_i$ oscillations in mouse zygotes and function of the inositol 1,4,5-trisphosphate receptor-1. Dev Biol 2004; 274(1):94-109.
25. Fissore RA, Dobrinsky JR, Balise JJ et al. Patterns of intracellular Ca^{2+} concentrations in fertilized bovine eggs. Biol Reprod 1992; 47(6):960-969.
26. Nakada K, Mizuno J, Shiraishi K et al. Initiation, persistence, and cessation of the series of intracellular Ca^{2+} responses during fertilization of bovine eggs. J Reprod Dev 1995; 41:77-84.
27. Stricker SA. Repetitive calcium waves induced by fertilization in the nemertean worm Cerebratulus lacteus. Dev Biol 1996; 176(2):243-263.
28. Carroll DJ, Ramarao CS, Mehlmann LM et al. Calcium release at fertilization in starfish eggs is mediated by phospholipase C gamma. J Cell Biol 1997; 138(6):1303-1311.
29. Giusti AF, Carroll DJ, Abassi YA et al. Evidence that a starfish egg Src family tyrosine kinase associates with PLC gamma1 SH2 domains at fertilization. Dev Biol 1999; 208(1):189-199.
30. Giusti AF, Carroll DJ, Abassi YA et al. Requirement of a Src family kinase for initiating calcium release at fertilization in starfish eggs. J Biol Chem 1999; 274(41):29318-29322.
31. Giusti AF, Xu W, Hinkle B et al. Evidence that fertilization activates starfish eggs by sequential activation of a Src-like kinase and phospholipase C gamma. J Biol Chem 2000; 275(22):16788-16794.
32. Sato K, Tokmakov AA, Iwasaki T et al. Tyrosine kinase-dependent activation of phospholipase Cgamma is required for calcium transient in Xenopus egg fertilization. Dev Biol 2000; 224(2):453-469.
33. Sato K, Tokmakov AA, He CL et al. Reconstitution of Src-dependent phospholipase Cgamma phosphorylation and transient calcium release by using membrane rafts and cell-free extracts from Xenopus eggs. J Biol Chem 2003; 278(40):38413-38420.
34. Shearer J, De Nadai C, Emily-Fenouil F et al. Role of phospholipase Cgamma at fertilization and during mitosis in sea urchin eggs and embryos. Development 1999; 126(10):2273-2284.
35. Tokmakov AA, Sato KI, Iwasaki T et al. Src kinase induces calcium release in Xenopus egg extracts via PLC gamma and IP3-dependent mechanism. Cell Calcium 2002; 32(1):11-20.
36. Ciapa B, Epel D. A rapid change in phosphorylation on tyrosine accompanies fertilization of sea urchin eggs. FEBS 1991; 295(1-3):167-170.
37. Runft LL, Carroll DJ, Gillett J et al. Identification of a starfish egg PLC gamma that regulates Ca^{2+} release at fertilization. Dev Biol 2004; 269(1):220-236.
38. Kurokawa M, Sato K, Smyth J et al. Evidence that activation of Src family kinase is not required for fertilization-associated $[Ca^{2+}]_i$ oscillations in mouse eggs. Reproduction 2004; 127(4):441-454.
39. Mehlmann LM, Carpenter G, Rhee SG et al. SH2 domain-mediated activation of phospholipase C gamma is not required to initiate Ca^{2+} release at fertilization of mouse eggs. Dev Biol 1998; 203(1):221-232.
40. Mehlmann LM, Chattopadhyay A, Carpenter G et al. Evidence that phospholipase C from the sperm is not responsible for initiating Ca^{2+} release at fertilization in mouse eggs. Dev Biol 2001; 236(2):492-501.
41. Dupont G, McGuinness OM, Johnson MH et al. Phospholipase C in mouse oocytes: Characterization of beta and gamma isoforms and their possible involvement in sperm-induced Ca^{2+} spiking. Biochem J 1996; 316(Pt 2):583-591.
42. Mehlmann LM, Jaffe LA. SH2 domain-mediated activation of an SRC family kinase is not required to initiate Ca^{2+} release at fertilization in mouse eggs. Reproduction 2005; 129(5):557-564.
43. Talmor A, Kinsey WH, Shalgi R. Expression and immunolocalization of p59c-fyn tyrosine kinase in rat eggs. Dev Biol 1998; 194(1):38-46.
44. Sette C, Paronetto MP, Barchi M et al. Tr-kit-induced resumption of the cell cycle in mouse eggs requires activation of a Src-like kinase. EMBO J 2002; 21(20):5386-5395.

45. Talmor-Cohen A, Tomashov-Matar R, Eliyahu E et al. Are Src family kinases involved in cell cycle resumption in rat eggs? Reproduction 2004; 127(4):455-463.
46. Miyazaki S. Inositol 1,4,5-trisphosphate-induced calcium release and guanine nucleotide-binding protein-mediated periodic calcium rises in golden hamster eggs. J Cell Biol 1988; 106(2):345-353.
47. Fissore RA, Robl JM. Mechanism of calcium oscillations in fertilized rabbit eggs. Dev Biol 1994; 166(2):634-642.
48. Moore GD, Ayabe T, Visconti PE et al. Roles of heterotrimeric and monomeric G proteins in sperm-induced activation of mouse eggs. Development 1994; 120(11):3313-3323.
49. Williams CJ, Mehlmann LM, Jaffe LA et al. Evidence that Gq family G proteins do not function in mouse egg activation at fertilization. Dev Biol 1998; 198(1):116-127.
50. Choi D, Lee E, Hwang S et al. The biological significance of phospholipase C beta 1 gene mutation in mouse sperm in the acrosome reaction, fertilization, and embryo development. J Assist Reprod Genet 2001; 18(5):305-310.
51. Parrington J, Swann K, Shevchenko VI et al. Calcium oscillations in mammalian eggs triggered by a soluble sperm protein. Nature 1996; 379(6563):364-368.
52. Swann K. A cytosolic sperm factor stimulates repetitive calcium increases and mimics fertilization in hamster eggs. Development 1990; 110(4):1295-302.
53. Stice SL, Robl JM. Activation of mammalian oocytes by a factor obtained from rabbit sperm. Mol Reprod Dev 1990; 25(3):272-280.
54. Wu H, He CL, Fissore RA. Injection of a porcine sperm factor triggers calcium oscillations in mouse oocytes and bovine eggs. Mol Reprod Dev 1997; 46(2):176-189.
55. Stricker SA. Intracellular injections of a soluble sperm factor trigger calcium oscillations and meiotic maturation in unfertilized oocytes of a marine worm. Dev Biol 1997; 186(2):185-201.
56. Dale B, DeFelice LJ, Ehrenstein G. Injection of a soluble sperm fraction into sea-urchin eggs triggers the cortical reaction. Experientia 1985; 41(8):1068-1070.
57. Palermo G, Joris H, Devroey P et al. Pregnancies after intracytoplasmic injection of single spermatozoon into an oocyte. Lancet 1992; 340(8810):17-18.
58. Tesarik J, Testart J. Treatment of sperm-injected human oocytes with Ca^{2+} ionophore supports the development of Ca^{2+} oscillations. Biol Reprod 1994; 51(3):385-391.
59. Kimura Y, Yanagimachi R, Kuretake S et al. Analysis of mouse oocyte activation suggests the involvement of sperm perinuclear material. Biol Reprod 1998; 58(6):1407-1415.
60. Perry AC, Wakayama T, Cooke IM et al. Mammalian oocyte activation by the synergistic action of discrete sperm head components: Induction of calcium transients and involvement of proteolysis. Dev Biol 2000; 217(2):386-393.
61. Jones KT, Matsuda M, Parrington J et al. Different Ca^{2+}-releasing abilities of sperm extracts compared with tissue extracts and phospholipase C isoforms in sea urchin egg homogenate and mouse eggs. Biochem J 2000; 346(Pt 3):743-749.
62. Wu H, Smyth J, Luzzi V et al. Sperm factor induces intracellular free calcium oscillations by stimulating the phosphoinositide pathway. Biol Reprod 2001; 64(5):1338-1349.
63. Perry AC, Wakayama T, Yanagimachi R. A novel trans-complementation assay suggests full mammalian oocyte activation is coordinately initiated by multiple, submembrane sperm components. Biol Reprod 1999; 60(3):747-755.
64. Sutovsky P, Oko R, Hewitson L et al. The removal of the sperm perinuclear theca and its association with the bovine oocyte surface during fertilization. Dev Biol 1997; 188(1):75-84.
65. Sutovsky P, Manandhar G, Wu A et al. Interactions of sperm perinuclear theca with the oocyte: Implications for oocyte activation, anti-polyspermy defense, and assisted reproduction. Microsc Res Tech 2003; 61(4):362-378.
66. Knott JG, Kurokawa M, Fissore RA. Release of the Ca^{2+} oscillation-inducing sperm factor during mouse fertilization. Dev Biol 2003; 260(2):536-547.
67. Kono T, Carroll J, Swann K et al. Nuclei from fertilized mouse embryos have calcium-releasing activity. Development 1995; 121(4):1123-1128.
68. Kono T, Jones KT, Bos-Mikich A et al. A cell cycle-associated change in Ca^{2+} releasing activity leads to the generation of Ca^{2+} transients in mouse embryos during the first mitotic division. J Cell Biol 1996; 132(5):915-923.
69. Fukami K, Nakao K, Inoue T et al. Requirement of phospholipase C delta4 for the zona pellucida-induced acrosome reaction. Science 2001; 292(5518):920-923.
70. Parrington J, Jones ML, Tunwell R et al. Phospholipase C isoforms in mammalian spermatozoa: Potential components of the sperm factor that causes Ca^{2+} release in eggs. Reproduction 2002; 123(1):31-39.
71. Saunders CM, Larman MG, Parrington J et al. PLC zeta: A sperm-specific trigger of $Ca^{(2+)}$ oscillations in eggs and embryo development. Development 2002; 129(15):3533-3544.

72. Cox LJ, Larman MG, Saunders CM et al. Sperm phospholipase C zeta from humans and cynomolgus monkeys triggers Ca^{2+} oscillations, activation and development of mouse oocytes. Reproduction 2002; 124(5):611-623.
73. Rhee SG. Regulation of phosphoinositide-specific phospholipase C. Annu Rev Biochem 2001; 70:281-312.
74. Essen LO, Perisic O, Lynch DE et al. A ternary metal binding site in the C2 domain of phosphoinositide-specific phospholipase C delta1. Biochemistry 1997; 36(10):2753-2762.
75. Williams RL, Katan M. Structural views of phosphoinositide-specific phospholipase C: Signalling the way ahead. Structure 1996; 4(12):1387-94.
76. Kouchi Z, Fukami K, Shikano T et al. Recombinant phospholipase C zeta has high Ca^{2+} sensitivity and induces Ca^{2+} oscillations in mouse eggs. J Biol Chem 2004; 279(11):10408-10412.
77. Fujimoto S, Yoshida N, Fukui T et al. Mammalian phospholipase C zeta induces oocyte activation from the sperm perinuclear matrix. Dev Biol 2004; 274(2):370-383.
78. Larman MG, Saunders CM, Carroll J et al. Cell cycle-dependent Ca^{2+} oscillations in mouse embryos are regulated by nuclear targeting of PLC zeta. J Cell Sci 2004; 117(Pt 12):2513-2521.
79. Rogers NT, Hobson E, Pickering S et al. Phospholipase Czeta causes Ca^{2+} oscillations and parthenogenetic activation of human oocytes. Reproduction 2004; 128(6):697-702
80. Malcuit C, Knott JG, He C et al. Fertilization and inositol 1,4,5-trisphosphate (IP3)-induced calcium release in type-1 inositol 1,4,5-trisphosphate receptor downregulated bovine eggs. Biol Reprod 2005; 73(1):2-13
81. Kouchi Z, Shikano T, Nakamura Y et al. The role of EF-hand domains and C2 domain in regulation of enzymatic activity of phospholipase Czeta. J Biol Chem 2005; 280(22):21015-21021.
82. Nomikos M, Blayney LM, Larman MG et al. Role of phospholipase C-zeta domains in Ca2+-dependent phosphatidylinositol 4,5-bisphosphate hydrolysis and cytoplasmic Ca2+ oscillations. J Biol Chem 2005; 280(35):31011-31018.
83. Yoda A, Oda S, Shikano T et al. Ca^{2+} oscillation-inducing phospholipase C zeta expressed in mouse eggs is accumulated to the pronucleus during egg activation. Dev Biol 2004; 268(2):245-257.
84. Lorca T, Cruzalegui FH, Fesquet D et al. Calmodulin-dependent protein kinase II mediates inactivation of MPF and CSF upon fertilization of Xenopus eggs. Nature 1993; 366(6452):270-273.
85. Rauh NR, Schmidt A, Bormann J et al. Calcium triggers exit from meiosis II by targeting the APC/C inhibitor XErp1 for degradation. Nature 2005; 437(7061):1048-1052.
86. Liu J, Maller JL. Calcium elevation at fertilization coordinates phosphorylation of XErp1/Emi2 by Plx1 and CaMK II to release metaphase arrest by cytostatic factor. Curr Biol 2005; 15(16):1458-1468.
87. King RW, Peters JM, Tugendreich S et al. A 20S complex containing CDC27 and CDC16 catalyzes the mitosis-specific conjugation of ubiquitin to cyclin B. Cell 1995; 81(2):279-288.
88. Yu H, King RW, Peters JM et al. Identification of a novel ubiquitin-conjugating enzyme involved in mitotic cyclin degradation. Curr Biol 1996; 6(4):455-466.
89. Nixon VL, Levasseur M, McDougall A et al. Ca^{2+} oscillations promote APC/C-dependent cyclin B1 degradation during metaphase arrest and completion of meiosis in fertilizing mouse eggs. Curr Biol 2002; 12(9):746-750.
90. Marangos P, Carroll J. The dynamics of cyclin B1 distribution during meiosis I in mouse oocytes. Reproduction 2004; 128(2):153-162.
91. Marangos P, Carroll J. Fertilization and InsP3-induced Ca^{2+} release stimulate a persistent increase in the rate of degradation of cyclin B1 specifically in mature mouse oocytes. Dev Biol 2004; 272(1):26-38.
92. Gautier J, Minshull J, Lohka M et al. Cyclin is a component of maturation-promoting factor from Xenopus. Cell 1990; 60(3):487-494.
93. Markoulaki S, Matson S, Abbott AL et al. Oscillatory CaMKII activity in mouse egg activation. Dev Biol 2003; 258(2):464-474.
94. Markoulaki S, Matson S, Ducibella T. Fertilization stimulates long-lasting oscillations of CaMKII activity in mouse eggs. Dev Biol 2004; 272(1):15-25.
95. Dupont G. Link between fertilization-induced Ca^{2+} oscillations and relief from metaphase II arrest in mammalian eggs: A model based on calmodulin-dependent kinase II activation. Biophys Chem 1998; 72(1-2):153-167.
96. Masui Y. A cytostatic factor in amphibian oocytes: Its extraction and partial characterization. J Exp Zool 1974; 187(1):141-147.
97. Ozil JP. The parthenogenetic development of rabbit oocytes after repetitive pulsatile electrical stimulation. Development 1990; 109(1):117-127.
98. Ozil JP, Huneau D. Activation of rabbit oocytes: The impact of the Ca^{2+} signal regime on development. Development 2001; 128(6):917-928.
99. Ozil JP, Markoulaki S, Toth S et al. Egg activation events are regulated by the duration of a sustained [Ca^{2+}]cyt signal in the mouse. Dev Biol 2005; 282(1):39-54.

100. Aoki F, Hara KT, Schultz RM. Acquisition of transcriptional competence in the 1-cell mouse embryo: Requirement for recruitment of maternal mRNAs. Mol Reprod Dev 2003; 64(3):270-274.
101. Tatone C, Delle Monache S et al. Possible role for Ca^{2+} calmodulin-dependent protein kinase II as an effector of the fertilization Ca^{2+} signal in mouse oocyte activation. Mol Hum Reprod 2002; 8(8):750-757.
102. De Koninck P, Schulman H. Sensitivity of CaM kinase II to the frequency of Ca^{2+} oscillations. Science 1998; 279(5348):227-230.
103. Madgwick S, Levasseur M, Jones KT. Calmodulin-dependent protein kinase II, and not protein kinase C, is sufficient for triggering cell-cycle resumption in mammalian eggs. J Cell Sci 2005; 118(Pt 17):3849-3859.
104. Atkins CM, Nozaki N, Shigeri Y et al. Cytoplasmic polyadenylation element binding protein-dependent protein synthesis is regulated by calcium/calmodulin-dependent protein kinase II. J Neurosci 2004; 24(22):5193-5201.
105. McAvey BA, Wortzman GB, Williams CJ et al. Involvement of calcium signaling and the actin cytoskeleton in the membrane block to polyspermy in mouse eggs. Biol Reprod 2002; 67(4):1342-1352.
106. Tutuncu L, Stein P, Ord TS et al. Calreticulin on the mouse egg surface mediates transmembrane signaling linked to cell cycle resumption. Dev Biol 2004; 270(1):246-260.
107. Wilmut I, Schnieke AE, McWhir J et al. Viable offspring derived from fetal and adult mammalian cells. Nature 1997; 385(6619):810-813.
108. Cibelli JB, Stice SL, Golueke PJ et al. Cloned transgenic calves produced from nonquiescent fetal fibroblasts. Science 1998; 280:1256-1258.
109. Baguisi A, Behboodi E, Melican DT et al. Production of goats by somatic cell nuclear transfer. Nat Biotech 1999; 17:456-461.
110. Ogura A, Inoue K, Takano K et al. Birth of mice after nuclear transfer by electrofusion using tail tip cells. Mol Reprod Dev 2000; 57(1):55-59.
111. Polejaeva IA, Chen SH, Vaught TD et al. Cloned pigs produced by nuclear transfer from adult somatic cells. Nature 2000; 407:86-90.
112. Chesne P, Adenot PG, Viglietta C et al. Cloned rabbits produced by nuclear transfer from adult somatic cells. Nat Biotechnol 2002; 20(4):366-369.
113. Shin T, Kraemer D, Pryor J et al. A cat cloned by nuclear transplantation. Nature 2002; 415(6874):859.
114. Woods GL, White KL, Vanderwall DK et al. A mule cloned from fetal cells by nuclear transfer. Science 2003; 301(5636):1063.
115. Galli C, Lagutina I, Crotti G et al. Pregnancy: A cloned horse born to its dam twin. Nature 2003; 424(6949):635.
116. Zhou Q, Renard JP, Le Friec G et al. Generation of fertile cloned rats by regulating oocyte activation. Science 2003; 302(5648):1179.
117. Cheek TR, McGuinness OM, Vincent C et al. Fertilisation and thimerosal stimulate similar calcium spiking patterns in mouse oocytes but by separate mechanisms. Development 1993; 119(1):179-189.
118. Wakayama T, Rodriguez I, Perry AC et al. Mice cloned from embryonic stem cells. Proc Natl Acad Sci 1999; 96(26):14984-14989.
119. Zhang D, Pan L, Yang LH et al. Strontium promotes calcium oscillations in mouse meiotic oocytes and early embryos through InsP3 receptors, and requires activation of phospholipase and the synergistic action of InsP3. Hum Reprod 2005; 20(11):3053-61.
120. Jellerette T, He CL, Wu H et al. Downregulation of the inositol 1,4,5-trisphosphate receptor in mouse eggs following fertilization or parthenogenetic activation. Dev Biol 2000; 223(2):238-250.
121. Knott JG, Poothapillai K, Wu H et al. Porcine sperm factor supports activation and development of bovine nuclear transfer embryos. Biol Reprod 2002; 66(4):1095-1103.
122. Campbell KHS, McWhir J, Ritchie WA et al. Sheep cloned by nuclear transfer from a cultured cell line. Nature 1996; 380:64-66.
123. Liu L, Ju JC, Yang X. Parthenogenetic development and protein patterns of newly matured bovine oocytes after chemical activation. Mol Reprod Dev 1998; 49(3):298-307.
124. Alberio R, Zakhartchenko V, Motlik J et al. Mammalian oocyte activation: Lessons from the sperm and implications for nuclear transfer. Int J Dev Biol 2001; 45(7):797-809.
125. Kumagai A, Dunphy WG. The cdc25 protein controls tyrosine dephosphorylation of the cdc2 protein in a cell-free system. Cell 1991; 64(5):903-914.
126. Liu L, Yang X. Interplay of maturation-promoting factor and mitogen-activated protein kinase inactivation during metaphase-to-interphase transition of activated bovine oocytes. Biol Reprod 1999; 61(1):1-7.

127. Zhang SC, Masui Y. Activation of Xenopus laevis eggs in the absence of intracellular Ca activity by the protein phosphorylation inhibitor, 6-dimethylaminopurine (6-DMAP). J Exp Zool 1992; 262(3):317-29.
128. Wells DN, Misica PM, Tervit HR. Production of cloned calves following nuclear transfer with cultured adult mural granulosa cells. Biol Reprod 1999; 60:996-1005.
129. Alexander B, Coppola G, Di Berardino D et al. The effect of 6-dimethylaminopurine (6-DMAP) and cycloheximide (CHX) on the development and chromosomal complement of sheep partheno-genetic and nuclear transfer embryos. Mol Reprod Dev 2006; 73(1):20-30.
130. Gray NS, Wodicka L, Thunnissen AM et al. Exploiting chemical libraries, structure, and genomics in the search for kinase inhibitors. Science 1998; 281(5376):533-538.
131. Shiga K, Fujita T, Hirose K et al. Production of calves by transfer of nuclei from cultured somatic cells obtained from Japanese black bulls. Theriogenology 1999; 52:527-535.
132. Soloy E, Kanka J, Viuff D et al. Time course of pronuclear deoxyribonucleic acid synthesis in parthenogenetically activated bovine oocytes. Biol Reprod 1997; 57(1):27-35.
133. Wang WH, Machaty Z, Abeydeera LR et al. Parthenogenetic activation of pig oocytes with calcium ionophore and the blcok to sperm penetration after activation. Biol Reprod 1998; 58:1357-1366.
134. Dean W, Santos F, Reik W. Epigenetic reprogramming in early mammalian development and following somatic nuclear transfer. Semin Cell Dev Biol 2003; 14(1):93-100.
135. Oback B, Wells DN. Cloning cattle. Cloning Stem Cells 2003; 5(4):243-256.
136. Campbell KH, Alberio R, Choi I et al. Cloning: Eight years after Dolly. Reprod Domest Anim 2005; 40(4):256-268.
137. Wrenzycki C, Wells D, Herrmann D et al. Nuclear transfer protocol affects messenger RNA expression patterns in cloned bovine blastocysts. Biol Reprod 2001; 65:309-317.
138. Latham KE. Mechanisms and control of embryonic genome activation in mammalian embryos. Int Rev Cytol 1999; 193:71-124.
139. Adenot PG, Mercier Y, Renard JP et al. Differential H4 acetylation of paternal and maternal chromatin precedes DNA replication and differential transcriptional activity in pronuclei of 1-cell mouse embryos. Development 1997; 124(22):4615-4625.
140. Santos F, Hendrich B, Reik W et al. Dynamic reprogramming of DNA methylation in the early mouse embryo. Dev Biol 2002; 241(1):172-182.
141. Barton SC, Arney KL, Shi W et al. Genome-wide methylation patterns in normal and uniparental early mouse embryos. Hum Mol Genet 2001; 10(26):2983-2987.
142. Sone Y, Ito M, Shirakawa H et al. Nuclear translocation of phospholipase C-zeta, an egg-activating factor, during early embryonic development. Biochem Biophys Res Commun 2005; 330(3):690-694.
143. Stein P, Worrad DM, Belyaev ND et al. Stage-dependent redistributions of acetylated histones in nuclei of the early preimplantation mouse embryo. Mol Reprod Dev 1997; 47(4):421-429.
144. Strahl BD, Allis CD. The language of covalent histone modifications. Nature 2000; 403(6765):41-45.
145. Fischle W, Wang Y, Allis CD. Binary switches and modification cassettes in histone biology and beyond. Nature 2003; 425(6957):475-479.
146. Sarmento OF, Digilio LC, Wang Y et al. Dynamic alterations of specific histone modifications during early murine development. J Cell Sci 2004; 117(Pt 19):4449-4459.
147. Wright PW, Bolling LC, Calvert ME et al. ePAD, an oocyte and early embryo-abundant peptidylarginine deiminase-like protein that localizes to egg cytoplasmic sheets. Dev Biol 2003; 256(1):73-88.
148. Wang Y, Wysocka J, Sayegh J et al. Human PAD4 regulates histone arginine methylation levels via demethylimination. Science 2004; 306(5694):279-283.
149. Vossenaar ER, Radstake TR, van der Heijden A et al. Expression and activity of citrullinating peptidylarginine deiminase enzymes in monocytes and macrophages. Ann Rheum Dis 2004; 63(4):373-381.

Index

A

Acetylation 16, 94, 98, 100, 125
Adult stem (AS) cell 39, 40
Assisted reproductive technique (ART) 72-74, 81

B

Behavior 55, 69, 72, 73, 77, 78, 81, 98, 100
Bimaternal inheritance 109
Biotechnology 56
Body composition 80
Body weight 56, 78-81

C

Ca^{2+} 117-126
CaMKII 119, 121, 122, 126
Cattle 1, 2, 4, 30-33, 36-42, 44-57, 65, 73, 74, 76, 78, 86, 93, 106-108, 110, 111, 123, 124
Cell cycle 3, 14, 20, 34-37, 41, 42, 58-60, 62, 63, 65, 66, 84, 86-91, 94, 100, 104, 117, 119, 122, 124
Centriole 4, 17, 19, 39, 58-64, 68, 69
Centrosome 3, 4, 39, 58-69
Chimerism 38, 112
Chromatin 2, 14, 16-19, 33, 36, 44, 84, 86, 88, 89, 94, 97-100
Chromosome 2, 4, 8, 15-19, 23, 33, 43, 44, 50, 59, 60, 62, 65, 66, 78, 94, 97-100
Clone 2, 4, 15-23, 32, 35, 37, 38, 40-44, 46-49, 72-81, 93, 97, 123, 124
Cloning 1, 2, 5, 14-19, 21, 23, 24, 30-45, 47-49, 64-66, 68, 72-74, 77, 78, 81, 84, 106, 113, 123, 125
Cloning efficiency 30, 32-45, 65
Cows 14, 47, 49, 62, 107, 108, 112, 113
Cumulus cell 20, 23, 37, 38, 40, 44, 47, 73-76, 78-80, 89, 94, 101, 109
Cyclin 65, 68, 94, 117, 119, 122, 124, 126
Cytoplasm 1-4, 33, 36, 39, 44, 45, 47, 60-63, 65, 67, 68, 73, 74, 84, 88-90, 94-99, 113, 114, 122
Cytoplast 33, 38, 44, 45, 67, 84, 87-90, 93-101, 113

D

Demethylation 5, 16, 37, 38, 99, 125
Dense fibrillar component (DFC) 86-90
Development 1, 3-5, 8, 15-21, 23, 32, 33, 35, 37, 38, 40, 42-47, 58, 62, 65, 67, 68, 72-77, 80, 81, 85, 86, 89-91, 94, 97-99, 103, 105, 111-113, 117, 120-125
Differentiation 32, 36-38, 40, 41, 43, 86
Dimethylaminopurine (DMAP) 90, 123, 124
DNA methylation 2-5, 16, 21, 23, 37, 42, 76, 94, 99, 125
Donor cell 2, 4, 5, 14-16, 18-21, 24, 30, 33-45, 58, 65-68, 74, 76, 78, 80, 84, 90, 91, 93, 99, 101, 105-114

E

Embryo 1-5, 14-24, 33, 35-39, 41-49, 58, 61, 63-69, 73-76, 84-91, 93, 94, 97, 99, 100, 103-108, 110-113, 117, 119, 120, 122-126
Embryonic cell nuclear transfer (ECNT) 36, 68, 84, 85, 87, 90
Embryonic stem (ES) cell 14, 19, 35, 36, 38, 40, 41, 43, 44, 47, 72, 74, 75, 77, 80, 81
Enucleation 15, 33, 39, 44, 65, 67, 87, 95-100
Epigenetic 3, 4, 16, 30, 32, 36-39, 42, 44, 47-49, 81, 84, 94, 98, 125, 126
Euchromatin 94

F

Fertilization 2, 4, 16, 17, 36, 37, 45, 58, 61-65, 67, 68, 73, 76, 84-86, 94, 98-100, 105, 112, 117-126
Fibrillar center (FCs) 86-91

G

Gametogenesis 5, 19, 32, 47, 58, 61, 64, 68, 81
Gene expression 1, 3, 6, 17, 19-21, 23, 42, 76, 84, 86, 91, 93, 94, 124-126
Genetic modification 2, 5, 6, 8, 49, 50
Granular component (GC) 86-90
Growth 6, 23, 46, 74, 90, 91, 97, 98, 104, 113